RAND NATIONAL DEFENSE RESEARCH INSTITUTE

Assessing Department of Defense Use of Data Analytics and Enabling Data Management to Improve Acquisition Outcomes

Philip S. Anton, Megan McKernan, Ken Munson, James G. Kallimani, Alexis Levedahl, Irv Blickstein, Jeffrey A. Drezner, Sydne Newberry

Prepared for the Office of the Secretary of Defense
Approved for public release; distribution unlimited

For more information on this publication, visit **www.rand.org/t/RR3136**

Library of Congress Cataloging-in-Publication Data is available for this publication.
ISBN: 978-1-9774-0326-1

Published by the RAND Corporation, Santa Monica, Calif.

Preface

Congress has heightened its interest in the U.S. Department of Defense's (DoD's) use of data analytics in acquisition decisionmaking. The Joint Explanatory Statement of the Committee of Conference (U.S. House of Representatives, 2016, pp. 1125–1126) accompanying the fiscal year 2017 National Defense Authorization Act (enacted on December 23, 2016) directed "the Secretary of Defense, not later than 1 year after the date of the enactment of this Act, to brief the Armed Services Committees of the Senate and House of Representatives on the use of data analysis, measurement, and other evaluation-related methods in DOD acquisition programs." The scope of Congress's inquiry includes the use of any type of relevant analytic and other evaluation-related methods for acquisition that utilize data, the availability and management of data (e.g., best practices for data collection, delivery, and availability), and opportunities to improve these aspects as they pertain to improving acquisition program management and decisionmaking.

As part of this effort, the Office of the Under Secretary of Defense for Acquisition and Sustainment (OUSD[A&S]) asked the RAND National Defense Research Institute to inform the secretary's briefing to the committees. This annotated briefing documents our analysis in support of that briefing.

This research was conducted from June 2018 to March 2019. This research should be of interest not only to the Office of the Secretary of Defense but also to stakeholders interested in the DoD's efforts to improve data collection, availability, and analysis to inform acquisition decisionmaking, as well as opportunities to extend those activities.

This research was sponsored by the Director, Acquisition Analytics and Policy, within the OUSD(A&S) and conducted within the Acquisition and Technology Policy Center of the RAND National Defense Research Institute, a federally funded research and development center sponsored by the Office of the Secretary of Defense, the Joint Staff, the Unified Combatant Commands, the Navy, the Marine Corps, the defense agencies, and the defense Intelligence Community.

For more information on RAND's Acquisition and Technology Policy Center, see www.rand.org/nsrd/ndri/centers/atp or contact the director (contact information is provided on the webpage).

Contents

Summary

Summary Overview Briefing

- *The following slides provide a summary overview of the study results.*
- *Detailed supporting slides follow.*

Congress has expressed concern over the past several years about the extent and currency of data analytics in the U.S. Department of Defense (DoD) and their use in the DoD's acquisition management and decisionmaking. In particular, the conference committee for the fiscal year (FY) 2017 National Defense Authorization Act (NDAA) noted "a widespread recognition that DoD does not sufficiently incorporate data into its acquisition-related learning and decision-making. . . . These policies are sometimes based on assumptions, and program reviews do not always sufficiently incorporate relevant data against which to evaluate success" (U.S. House of Representatives, 2016, p. 1125). Therefore, the conference committee directed the Secretary of Defense to "brief the Armed Services Committees of the Senate and House of Representatives on the use of data analysis, measurement, and other evaluation-related methods in DoD acquisition programs" (U.S. House of Representatives, 2016, p. 1126). The scope of Congress's inquiry includes the use of any type of relevant analytic and other evaluation-related methods for acquisition that utilize data, the enabling of data management (e.g., best practices for data collection, delivery, and availability), and opportunities to improve these aspects as they pertain to improving acquisition program management and decisionmaking.

The following slides provide an extended summary of the results from the study.

Congressional Tasking

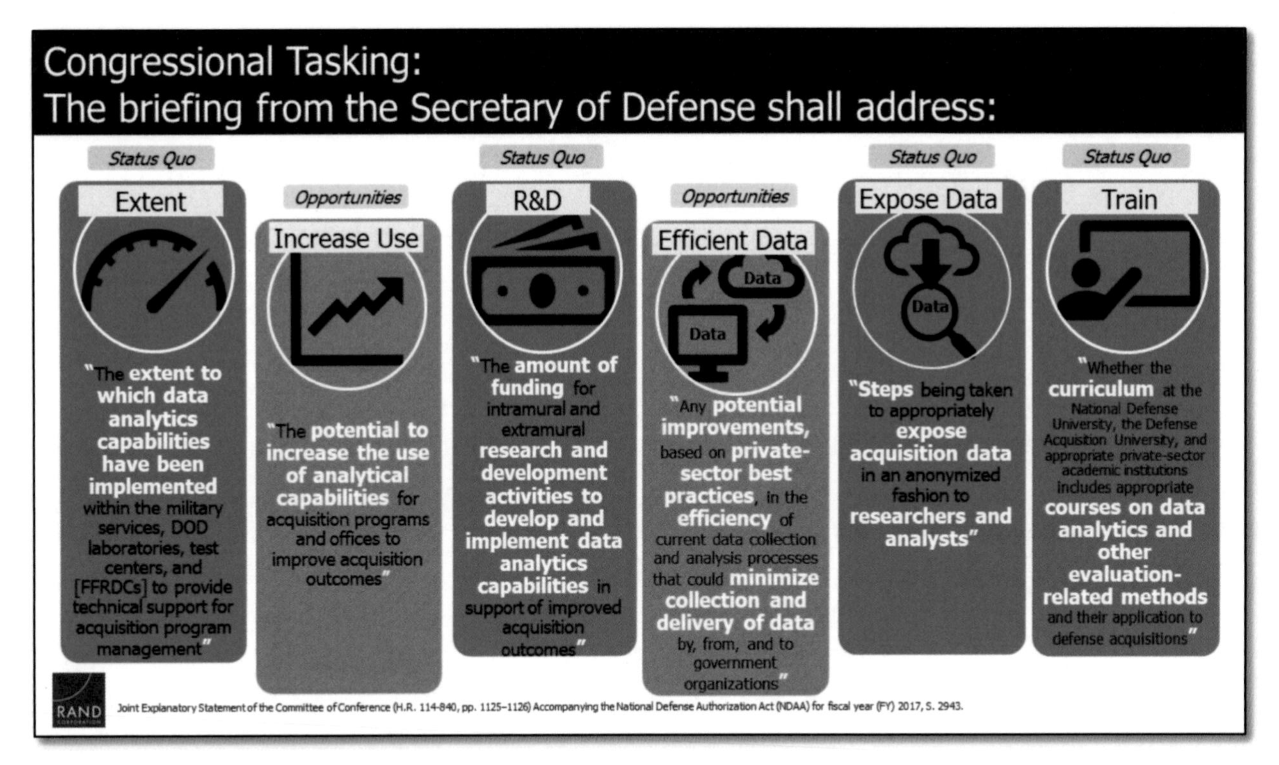

SOURCE: U.S. House of Representatives, 2016, p. 1126.
NOTE: FFRDC = federally funded research and development center; R&D = research and development.

Congress directed that the briefing address six topics: four topics about the status quo and two about opportunities (U.S. House of Representatives, 2016, p. 1126). We framed these as the following research questions that align directly with Congress's six topics:

1. **Current extent of implementation**: What is the extent to which data analytics capabilities have been implemented across the DoD to provide technical support for acquisition program management?
2. **Potential to increase use**: What is the potential to increase the use of analytical capabilities?
3. **Current R&D funding**: What is the current amount of R&D funding for acquisition data analytics capabilities?
4. **Potential improvements in data collection and analysis**: What private-sector best practices could be leveraged to minimize the collection and delivery of data?
5. **Current data-sharing efforts**: What steps are being taken to share anonymized acquisition data to researchers and analysts?
6. **Current data analytics training**: Do the curricula at defense acquisition workforce (AWF) training institutions include appropriate courses on applied and general data analytics and other evaluation-related methods?

This briefing summarizes our analysis on these six topics and associated questions and offers perspectives to help Congress understand the larger context and realities of the findings, as well as opportunities and suggestions for Congress and the DoD.

For this analysis, we considered *acquisition* as encompassing "[t]he conceptualization, initiation, design, development, test, contracting, production, deployment, integrated product support, modification, and disposal of weapons and other systems, supplies, or services (including construction) to satisfy DoD needs, intended for use in, or in support of, military missions" (Defense Acquisition University [DAU], 2017a). Based on the conference report elements, we used the following as our definition of *data analytics*: data analysis, measurement, and other evaluation-related methods (i.e., techniques to assess and analyze data) to inform acquisition decisions, policymaking, program management, evaluation, and learning.

Research Approach, Scope, and Limitations

Research approach addresses the breadth of Congressional topics: data, analysis, and improvement opportunities

- Inputs:
  - Discussions with select stakeholders
  - Existing literature, public budget exhibits, policies, legislation, training course catalogs and curricula; inventories of relative IT systems
  - Experience, knowledge, and judgement of authors, as noted
- Study techniques
  - Functional decomposition and mapping; workforce cost analysis; budget and solicitation analysis; literature review and analysis; analysis of industry best practices; analysis of DoD data management and analytic activities.
- Assumptions
  - Data analytics are a means for improving decisions
    - *We lay out analytics that support specific decisions, but we generally do not directly study cause-and-effect*
  - Advanced business analytics have some DoD applications
    - *Found some application and explorations, but further applications require continued investigation*

Approach

Our study employed a variety of approaches to answer Congress's six questions. Various types of information and expertise were employed, as indicated on the slide. In particular, we used the decomposition and review of acquisition into functions as a means of understanding how data and associated analytics and other evaluation-related methods are used to support acquisition decisions. The breadth of sources and their seemingly diverse nature reflect the breadth of the questions and the fact that activities, capabilities, and budgets are often not tracked by data analytics and enabling data feeds.

The breadth of questions and limits of data also drove us to use a wide range of analyses to gain some perspective on the answers. These are outlined briefly on the slide.

Finally, we should note that we generally assumed that there is a relationship between data analytics and improved decisionmaking. We lay out such capabilities but generally did not focus on cause and effect. We also assumed that (rather than studied in depth) there may be uses for recent advanced analytics (particularly those used by the commercial business sector) in the DoD; we found some applications, but, given the breadth of Congress's topics, we did not focus solely on such data analytics.

Research approach embraces the breadth of congressional interest with some limitations on study scope and depth

- Broad scope driven by inclusive definitions of "acquisition" and "analytics" (see next slide) as well as the broad nature of Congress's questions
- Did not assess what specific acquisition data or data analytics are needed
- A survey (a "data call") was proposed to solicit specific examples of data analytics underway in the DoD acquisition community, but it was deemed infeasible within available time and resources
  - In lieu of a survey, primary inputs include discussions with select stakeholders; existing literature; public budget exhibits; policies and legislation; training catalogs and curricula; and inventories of relevant IT information systems
- Experience, knowledge and author judgment was used to synthesize information and fill gaps in primary inputs, published research and other secondary data
  - Where author judgment is applied, it is noted

Scope and Limitations

In assessing the extent of data analytics capabilities, we explore not only what data, analytic capabilities, and analytic results exist in defense acquisition but also how they are used.

It is important to note that the definitions of *acquisition* and *data analytics* scope our study. The broad definition of acquisition means that we did not limit our study to development and procurement but included subsequent activity, such as sustainment. It also is important to note that, in focusing on acquisition, we did not address analytics for related areas that affect acquisition, including force planning, requirements, budgeting and resourcing, intelligence, and military operations.

Our definition of *data analytics* implies that we did not limit our study to what might be termed *advanced* analytics but instead examined what Congress asked about: the full range of data analytics and other evaluation-related methods. Advanced analytics for acquisition could be deemed, for example, the application of "big data," machine learning, and predictive analytics to various acquisition data. Advanced analytics could also be more-mature developments, such as transforming systems engineering to complex model-based data analysis and design systems. We did, though, seek to assess what advanced techniques are being pursued and what kind of applications they may be appropriate for. We also looked at examples of data analytics as a continuum to illustrate any advancement made by the DoD and how simpler analytics can still be useful, but did not try to define a strict set of levels in that continuum.

Although we examined the breadth of data analytics and enabling data for major acquisition decisions (e.g., Milestone B) and in information systems that archive and make available various data, we did not assess in detail what specific data analytics and enabling data are needed and not needed. For example, we did not assess the specific minimum data needed at various levels in the acquisition chain of command for analysis and oversight.

For inputs to our study, a "data call" was initially proposed as a mechanism to collect information about Congress's six questions, but it was deemed infeasible and would likely not have produced consistent data that could be assessed DoD-wide. Also, instead of asking DoD components to assess available information on data analytics, we and our sponsors decided to examine those data ourselves to both reduce burden on the DoD components and seek a more consistent methodology across the information. Author experience, knowledge, and judgment were employed in our study to assess and synthesize available information and fill gaps in primary and secondary inputs. Research for this study was conducted by the following team members:

- Philip S. Anton is a senior information scientist who conducts research on acquisition and sustainment policy, cybersecurity, emerging technologies, technology foresight, process performance measurement and efficiency, data science and analytics, aeronautics test infrastructure, and military modeling and simulation.
- Megan McKernan is a senior defense researcher with more than 14 years of experience conducting DoD acquisition analyses. She is co-leading research examining DoD acquisition data management.
- Ken Munson is a senior management scientist specializing in acquisition, logistics, budget, requirements, and policy analysis.
- James G. Kallimani is a senior technical analyst with an engineering and data analytics background, specializing in problem-solving and creative thinking for public policy, national security, defense, and government.
- Alexis Levedahl is a research assistant with experience in emerging technologies, acquisition, and requirement development.
- Irv Blickstein is a senior engineer with 50 years of experience in the field of defense analysis and management, with a specialty in planning, programming, and budgeting, as well as acquisition.

- Jeffrey A. Drezner is a senior policy researcher who has conducted policy analysis on a wide range of issues, including energy R&D planning and program management, best practices in environmental management, analyses of cost and schedule outcomes in complex system development programs, aerospace industrial policy, defense acquisition policy and reform, and local emergency response.

Finally, there are other research questions in the area of data analytics and enabling data that we could have studied but are out of scope (e.g., what data are most valuable, and how can we understand costs and benefits; how can we assess the appropriateness of data usage; what specific approaches can be used to counter the culture of not sharing data; what is the optimal mix of advanced data analytics, commercial off-the-shelf (COTS) tools, and simpler analytic approaches; is the extent of data analytics capabilities sufficient or appropriate [how much spending is appropriate for data analytics—against what baseline and benchmarks]; and what is the quality of the data analytics courses at defense training institutions?). Some of these research topics are possible next steps.

Definitions

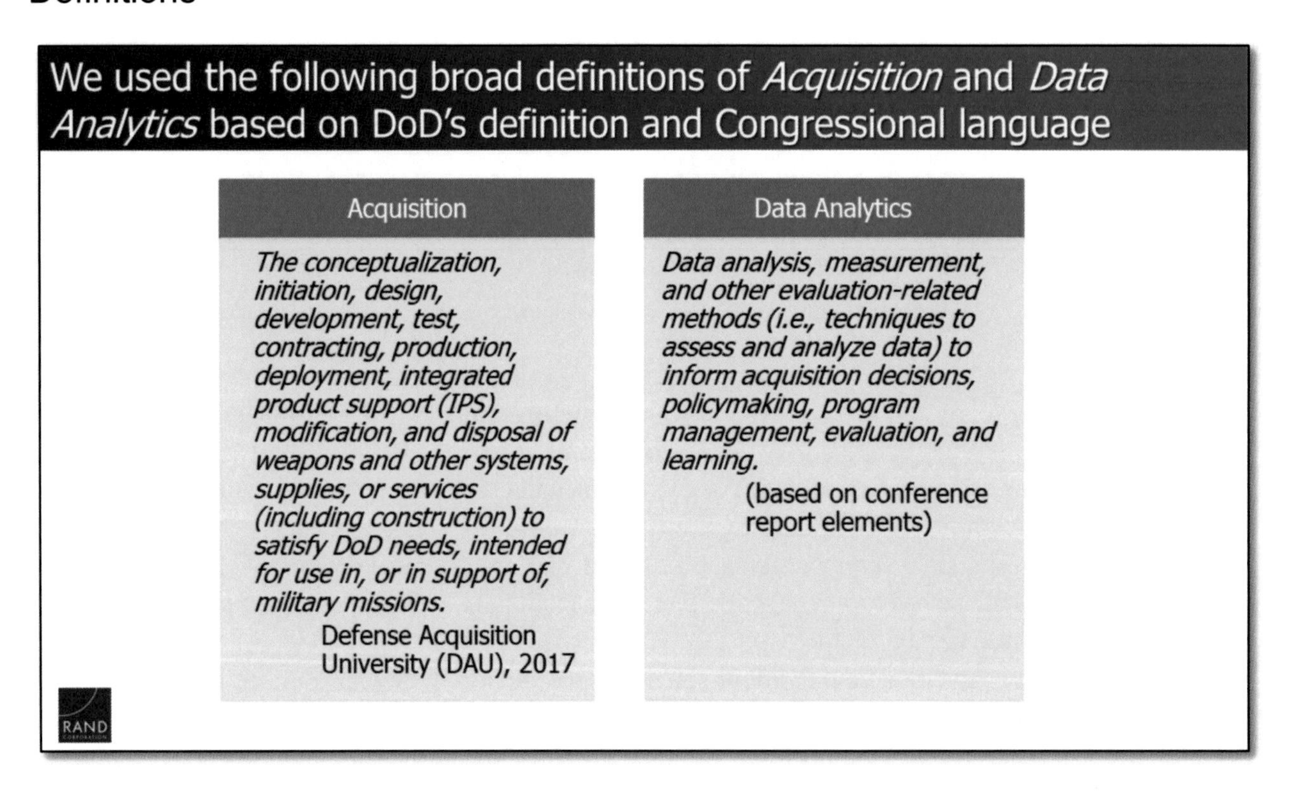

For our analysis, we used the DoD's official definition of *acquisition*, shown on the slide, as published by DAU (2017a). For what constitutes *data analytics* for our study, we constructed the definition shown on the slide based on our general reading of the conference report (U.S. House of Representatives, 2016, pp. 1125–1126) and our own knowledge of the field.

Bottom-Line Conclusions About Data Analytics

Bottom-Line Conclusions (in prose)

Acquisition data analytics is extensive:

- DoD Data and analytics supports acquisition decision-making through a wide range of acquisition functions
  - Those functions include fields used for decades: e.g., engineering, test and evaluation, cost estimating, auditing, and many others
  - The data analytics employed by these applied functions range from simple empirical methods to advanced engineering and estimating approaches
- It is difficult to precisely measure the magnitude of annual investment in data analytics, but rough macro-level, parametric estimates are in the billions of dollars.
- Cases of some major acquisition program decisions that went counter to the extant data and analysis appear to have been made for other strategic reasons or drivers—not for broad lack of data analytics

Still, there is room for improvement:

- Applying advanced commercial business data analytics to DoD acquisition program data is limited and mostly in the exploratory phase
- Barriers to expanded data and analytics include:
  - Some cases of inefficient collection and lack of sharing data due to cultural, security, and investment issues
  - Limited analytic desktop software for general staff
  - Workforce trained on job-specific data analytics rather than broader analytic skills
- Progress has been made to:
  - Apply commercial best practices to efficient data management and sharing
  - Add commercial business intelligence tools (e.g., Tableau, QlikSense) to information systems to facilitate data processing, analysis, and visualization
  - Offer new courses in data analytics at acquisition training institutions
- Continued investments and progress is needed
  - DoD appears to recognize this based on continued (albeit constrained) investments

Our study provides the following concise bottom-line conclusions.

Data Analytics Support a Broad Range of Functions

Data analytics in the DoD support a broad range of acquisition functions. The DoD is using a mix of advanced data analytics, COTS tools, and simpler approaches. What matters is using appropriate techniques to inform acquisition decisions, not the label of whether a particular technique is more advanced or has been used for a long time. Whether the mix employed by the DoD is optimal is beyond the scope of this study, but we recognize that the DoD has many useful applications and is continuing to explore new techniques to ascertain their appropriateness, utility, and cost.

Data Analytics as a Factor in Decisionmaking

Data analytics are used to inform DoD decisionmaking throughout the acquisition community and across acquisition functions. These analyses inform acquisition management, insight, oversight, execution, and decisions at all levels, including program managers, contracting officers, engineers, auditors, oversight executives, and DoD leadership. However, data analysis may or may not be equally weighted against other considerations by decisionmakers. At times, there may be other priorities (e.g., politics, optics) that will override analysis. Evidence shows that data analytics results are sometimes not directly followed by decisionmakers for other reasons, and some of these decisions ended up being problematic in hindsight (Director of Performance Assessments and Root Cause Analyses [PARCA], 2010a, 2010b, 2011a, 2011b, 2011c, 2013, 2016, 2017, 2018; Aronin et al., 2011; Blickstein et al., 2011, 2012; Michealson, 2012; Pernin et al., 2012; Work, 2013; Under Secretary of Defense for Acquisition, Technology, and Logistics, 2016a; O'Rourke, 2018). Two reasons among many could be that the analysis was wrong or not persuasive, so decisionmakers chose another course of action. This illustrates the difficulty in trying to tie analysis to decisions, but we were able to review some well-documented problematic programs to see whether their Nunn-McCurdy root-cause analyses indicated where information was available but was not fully acted on by leadership. According to the Office of PARCA, such cases are limited to about one-quarter of problematic programs (Under Secretary of Defense for Acquisition, Technology, and Logistics, 2016a) and involves mitigating incentives, strategic risk taking, and natural limitations when human judgment is involved.

The DoD Is Advancing Its Capabilities, but More Remains to Be Done

The DoD has made progress in improving its data and analysis capabilities, including implementations of the latest commercial analytic tools on top of open information systems. The DoD is also pursuing commercial best practices in data collection and management as it evolves its information systems in its main lines of acquisition functions. However, more remains to be done. The DoD is continuing to pursue investments and improvements along many of these best practices.

Remaining Challenges

Although data sharing has seen some improvement, remaining barriers include the following:

- a culture of data restriction
- security concerns
- lack of trust that data will be used appropriately (e.g., historical tendencies to staff invasiveness, micromanagement, and what economists call "moral hazards" [actions with side consequences not borne by the actor])
- the burden of data reporting.

Also, hard questions still require analysis specialists who understand the acquisition domain in question and can involve extensive data collection, cleaning, and management. As with the private sector, the pool of experts who understand both acquisition and data analytics is limited. In addition, recent breakthroughs in advanced analytics encourage the idea that they can easily be applied broadly to the DoD's problems, but they might not apply for various reasons. For example, we do not expect that clustering and machine-learning tools can be directly applied to predict program performance, since programs fail for different reasons, making this something different from a clustering problem. Nevertheless, research is being funded at universities and laboratories in a search for useful applications.

Need to Scope Expectations for Data Analytics

Finally, one should scope expectations of what data analytics can do for acquisition outcomes. The (albeit spectacular) recent commercial breakthroughs in data analytics do not always apply to the DoD, with its different mission and complicated incentives and constraints. Also, data analytics span a range of complexity. Easier efforts can involve relatively simple analysis and visualization of readily available data (e.g., business intelligence data). Deeper, more-complicated insights require other data over long periods that must be cultivated, managed, cleaned, and archived, requiring extensive and ongoing work. They cannot be constructed in a few weeks but rather require strategic planning and investing. We must also remember that data analysis is important for informing management and decisionmaking, but it cannot eliminate the risks inherent in acquisition—especially developmental defense acquisition that is rushing to deliver beyond-state-of-the-art capabilities to warfighters to gain a (temporary) competitive technical advantage. There is no magic bullet, but there is a range of options that are quite useful and can become more useful. The DoD is on that path but needs continued support and rational expectations. Finally, it is also implicitly assumed that better analysis means better outcomes. An important goal is to have better-informed decisionmakers who will make better decisions, resulting in better outcomes; therefore, the workforce part of this challenge is particularly important.

Question 1: Extent of Analytic Capabilities

Q1. To what extent have data analytics capabilities been implemented?

Findings:

- The breadth of data analytics across acquisition functions gives a measure of extent *(see next slide)*
- Spending indicates extent:
  - One estimate: $11–15 billion/year on analytic-related workforce
  - About $3 billion/year on IT systems for acquisition (FY 2019 budget exhibits)
    - *Major IT systems supporting acquisition processes and data collection. Some have analytic layers.*
- Data analyses are often available, but final decisions reflect risks and other equities

Question 1 asks about the extent to which "data analytics capabilities have been implemented within the military services, DOD laboratories, test centers, and Federally Funded Research and Development Centers to provide technical support for acquisition program management" (U.S. House of Representatives, 2016, p. 1125).

Findings

A wide breadth of data analytics exists across acquisition functions. DoD data analytics in support of acquisition program management and decisionmaking encompass a large array of data analytics applications, including market research, cost estimating, risk analysis, engineering, test and evaluation (T&E), security, supply chain, contracting, production, auditing, and sustainment (see the following slide). This broad range of analytic capabilities are reflected in the large number of informational documents that summarize data analysis at key acquisition decision points such as Milestone B (see, for example, the informational elements outlined in Table 2 of Department of Defense Instruction 5000.02, 2017). Thus, this extent includes much more than the typical programmatic execution data and information included in, for example, the Selected Acquisition Reports to Congress.

Spending indicates extent. Separately measuring the extent of analytic capabilities supporting acquisition is difficult, given that they are not accounted for as such in the DoD's workforce and operation budgets. Still, some parametric estimates are possible. Cost estimating, engineering, T&E, production and quality management, and auditing functions are predominantly data analytic in nature. One estimate indicates that the DoD spends about $13 billion per year on these capabilities (plus or minus $2 billion per year), using the average wage levels reported by the U.S. Bureau of Labor Statistics (adapted from Anton et al., forthcoming).

In terms of major IT systems supporting acquisition (not desktop computing), the budget exhibits include about $3 billion per year (about $0.5 billion for acquisition systems and about $2.5 billion for logistics and supply-chain systems). These information systems have a mix of acquisition process support, data collection and archiving, and data analytics layers; this helps to measure the underlying enabling capability that ultimately informs acquisition decisions during execution, management, and oversight.

Over the past decade or so, the DoD has been evolving its information systems from simpler isolated archives to systems with open data-sharing and analytic layers. These analytic tools go beyond dashboards by enabling both custom and predefined analyses.

Data analytics contribute to major program decisions, along with other considerations. To gain some perspective on the net effect of these functional data analytics on major program decisions, we reviewed 13 historical high-visibility programs (including those identified by the Office of Performance Assessments and Root Cause Analyses as involving decisionmakers' "failure to act on information") and whether data analysis was available to inform decisions. Generally, the literature shows that data analysis was available for decisionmakers, but the data analysis might have been outweighed by other factors deemed more important (e.g., political ones, mission impact). Examples include the following:

- Littoral Combat Ship (LCS; Milestone B in 2011): The cost community knew that the cost of the desired ships would be in the $500 million range (see, for example, Work, 2013). Driving for lower-cost ships was largely an initial leadership decision. Requirements were out of line with cost and weight realities, but analysis was available.
- Future Combat System (FCS; Milestone B in 2003): Leadership used the weight requirement to try to drive vehicle innovation (see, for example, Pernin et al., 2012). The cost and engineering analytic communities already forecast that such results were not possible. Requirements were out of line with cost and weight realities, respectively, but analysis was available. The ability to link ground and flying assets was not yet proven during the FCS project's time frame. The problem was intractable, and there were other earlier analytic shortfalls.

Some cases revealed shortfalls in capability, but the shortfalls appear to be more in staffing and experience than in whole-scale capability. Examples include the following:

- GPS Next-Generation Operational Control System (OCX; Milestone B in 2012): Although the problems were mostly from schedule risks, some workforce analytic shortfalls contributed to cost growth. The three root causes were an unrealistic schedule driven by the need to sustain the GPS constellation, inexperience driving underestimation of the cost to fully implement new information assurance instructions, and systems engineering issues (insufficient software engineering expertise and tenure in the program office; quality FFRDC analysis available but not sufficiently listened to; and insufficient contractor requirement decomposition and configuration management) (see Director of PARCA, 2016).

Also, some problems arise from changing situations rather than shortfalls in data analytics capabilities. One example is the following:

- DDG-1000 (Milestone B in 2005): The cost growth for the DDG-1000 is largely attributed to quantity reductions (Director of PARCA, 2010a; Blickstein et al., 2011), driven by a shift in mission emphasis (O'Rourke, 2018). Also, the "RDT&E [research, development, test, and evaluation] growth is largely explained by the addition of subsystems to the scope of work," and quantity adjustments of the original baseline estimates put unit costs below the baseline (Blickstein et al., 2011). Finally, additional (higher) cost estimates informed the baseline decision (Blickstein et al., 2011). Thus, despite some additional smaller issues, significant analysis supported the milestone decision, and external quantity changes were the major problem for cost growth.

How often do these issues occur? Finally, the frequency counts of problems give some perspective on how often these issues arise. For example, one measure of problematic programs is when programs breach the statutory Nunn-McCurdy cost-growth thresholds. The Office of the Secretary of Defense's (OSD's) independent analysis on such breaches since 2010 indicated whether the root causes were decisions that ran counter to available data analytics results. These root-cause analyses found that three of 16 Nunn-McCurdy breach programs and three of five discretionary root-cause analyses involved other reasons that led to decisions that ran counter to available data analytics results (see Under Secretary of Defense for Acquisition, Technology, and Logistics, 2016a, p. 28). Some breach cases involved a shortfall in capabilities (e.g., insufficient systems engineering staffing), but this situation does not appear to be a DoD-wide shortfall. We also note that 26 percent (six of 23) of independent root-cause analyses (breach and discretionary) performed by PARCA involved cases in which information was available but was not fully acted on by leadership (Under Secretary of Defense for Acquisition, Technology, and Logistics 2016a; Director of PARCA, 2018).

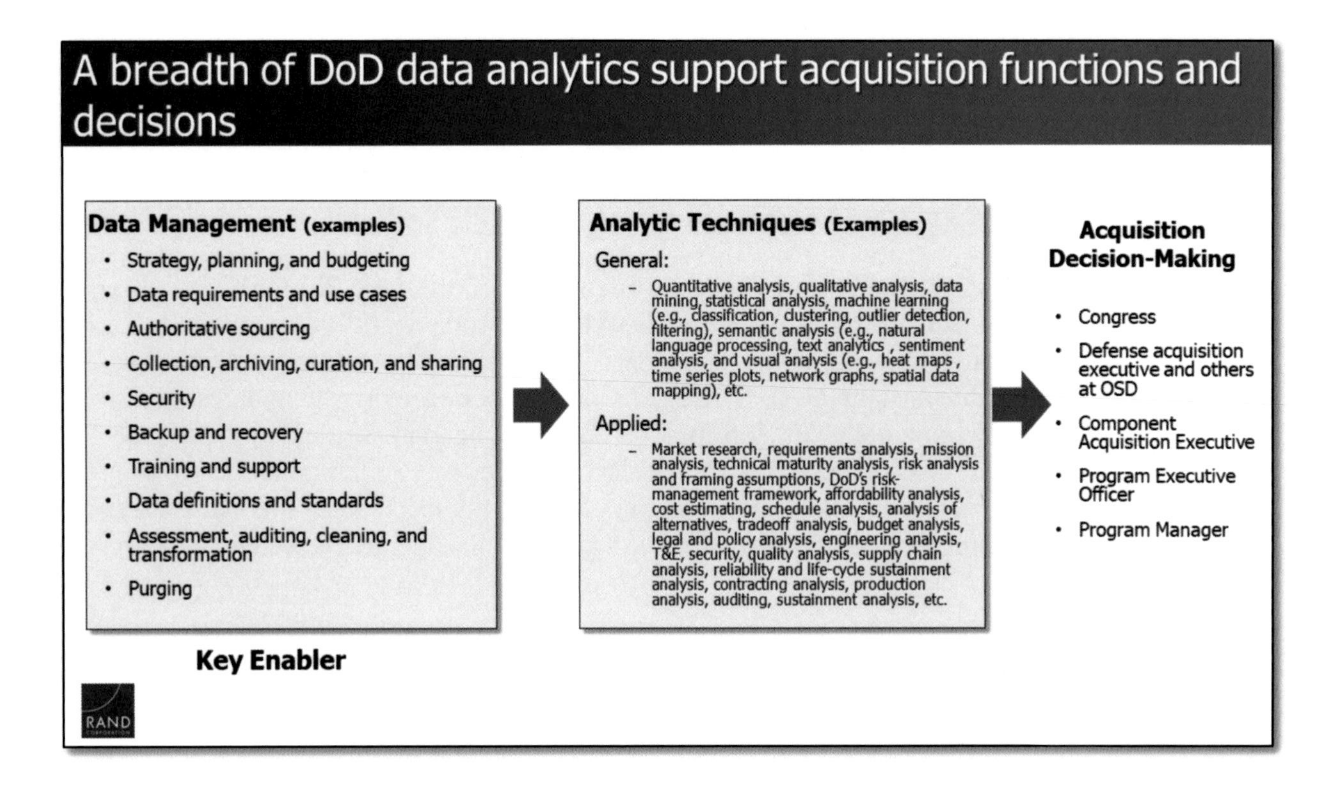

DoD data analytics that support acquisition functions and decisions across the acquisition chain of command (the program manager, program executive officer, and service acquisition executive, or the defense acquisition executive) and for other stakeholders (e.g., within the Office of the Secretary of Defense and in Congress) are broad. We mapped these data and analytic techniques, as executed for major acquisition milestone decisions, to understand how these data and analytic capabilities are used for major decisions.

Data management is necessary for doing data analytics. A key enabler of analysis is the data and associated management. Besides the obvious point that analysis cannot be conducted without data, this exposé on data management and governance helps to illustrate that significant activity and investments are required to obtain and provide data of sufficient quality for analysis. Even advanced commercial data analytics efforts and best practices often involve significant data collection, cleaning, and preprocessing. Challenges include ensuring common data definitions to enable cross-organizational data analysis. In some cases, existing accounting standards and charts of account already provide such standardization for basic business-intelligence analysis, but in other domains, commonality requires governance and active management. Sometimes data do not have to be standardized across all entities if the local definitions are available and can be transformed into a common standard, but that ability requires at least some minimal level of data definitions. Finally, some advanced analytic tools can ingest unstructured data, but such lack of structure imposes an added difficulty to the analysis, and those applications may be limited. Generally, we found many cases in which the DoD implemented these commercial best practices in data management and governance, but the level of maturity in these practices varied considerably across the DoD acquisition community.

Analytic techniques include those that are generally applicable and those used in applying data analytics to acquisition. These data then enable analysis, which can be grouped into two categories. *General techniques* are a wide variety of approaches that are independent of the specific application area. *Applied techniques* are specific (in our case) to various acquisition functions and analysis. The list in the previous slide helps illustrate what it means to *use* data analytics to inform acquisition decisions. *Use* is often through functions that are related to various aspects of acquisition, not just in terms of the types of generic tools employed (e.g., Excel, Tableau, Qlik Sense, R, Stata, "big data," and text understanding). In other words, an application is necessary when employing a tool, if that tool will inform a decision, and this list helps to illustrate the breadth of that application for acquisition.

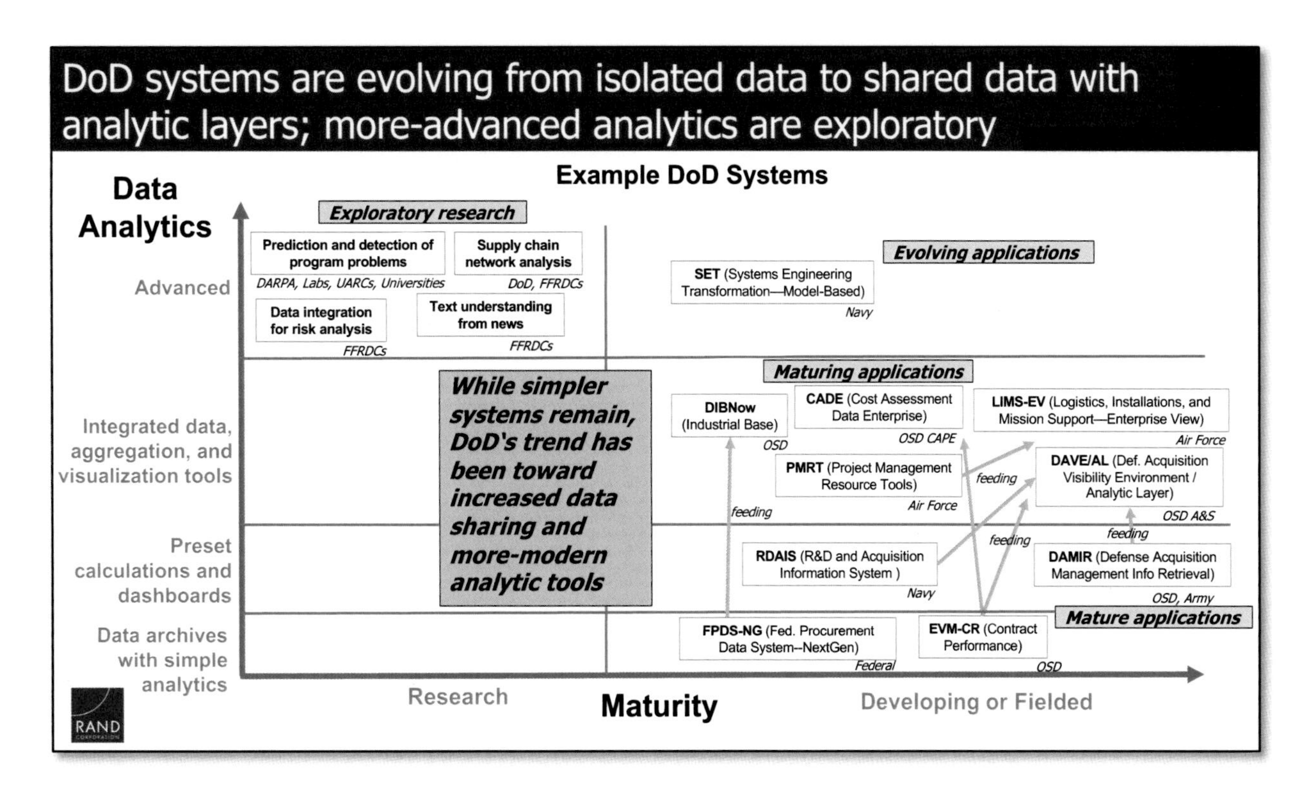

The DoD's information systems have evolved from isolated data archives to shared data systems that feed analytic information systems.

Data analytics range from simple to more advanced. The y-axis in the slide shows that analytics range from modern COTS analytic and visualization tools that are replacing predefined analytics and dashboards to more-advanced and more-specialized analytics. COTS analytic tools (e.g., business intelligence tools, such as Qlik Sense and Tableau) are installed on more systems across acquisition functions, including programmatic, cost, contract, logistics, and sustainment. Advanced analytics for acquisition range from exploratory research (where the DoD is trying to apply big data, machine learning, and predictive analytics to various acquisition data) to more-mature developments, such as transforming systems engineering to complex model-based data analysis and design systems.

Data maturity is lower for more-advanced applications. Acquisition data lay a foundation for the DoD's decisionmaking, execution, management, insight, and oversight of the weapon-system acquisition portfolio. Acquisition data help inform, monitor, and achieve many objectives (e.g., promoting transparency in spending, conducting analyses for improved decisionmaking, and archiving decisions). Acquisition data can be structured (that is, immediately identified within an electronic structure, such as a relational database) or unstructured (that is, not in fixed locations but often in free-form text). Such data are collected for a variety of statutory and regulatory requirements and at all levels, from program offices in the military departments to OSD, in both centralized and decentralized locations. The DoD also uses data from other federal information systems outside the department. The data themselves differ in time frame (with some dating back more than five decades), while the information systems that contain these data have different hardware, software, and interfaces. Technological improvements have helped the DoD improve data collection efficiency, quality, aggregation, ease of access and use, archiving, and analysis, among other characteristics; however, many information systems are difficult for users to navigate effectively and can take years to fully understand. Most systems are built for reporting, not analysis (McKernan et al., 2017).

The DoD has established many centralized information systems during the past 20 years in a variety of functional business areas, including R&D, requirements, budgeting, contracting, program cost, human capital, and acquisition oversight to make data more available for management, analysis, and decisionmaking. Acquisition decisionmakers use these data to answer a wide variety of questions related to defense acquisition. Although the data sets vary considerably, many share two strengths: standardization and collection of acquisition-related information in one place where data can be input, accessed, and analyzed by those needing it. The DoD is very large, and acquisitions are accomplished by many different organizations. A centralized system with consistent formats helps improve certain aspects of data consistency and quality. Also, analytic tools are layered on top of the data portions of some of the IT system architecture with varying levels of maturity (McKernan et al., 2017). These analytic tools go beyond dashboards by enabling both custom and predefined analyses.

The previous slide provides a simplistic visualization of examples of various centralized information systems and whether the data structure in those information systems is structured, unstructured, or both. The y-axis shows the capabilities of those information systems, ranging from simple storage to advanced analytics. This assessment is not meant to be a judgment of whether the DoD's capabilities are sufficient. Rather, it illustrates the range of capabilities across acquisition functions. It also illustrates that the DoD has established a variety of information systems over the past several years or matured prior capabilities in the military departments and OSD to provide data and more-advanced business intelligence analytic layers. For example, OSD has moved to the Defense Acquisition Visibility Environment (DAVE) for acquisition program information, which contains the recently added "analytic layer" for advanced data scientists. Cost analysts in OSD have also matured their analytic capabilities with the Cost Assessment Data Enterprise (CADE). The Air Force has moved toward an advanced business intelligence capability for program data with the Project Management Resource Tools (PMRT) information system, while the Army and the Navy are both considering options to improve the analytic capabilities for their AWFs. The industrial base community has likewise added more-advanced capabilities with DIBNow. Other efforts that include big-data mining and integration and predictive analytics are being pursued through work in the Defense Advanced Research Projects Agency (DARPA), DoD laboratories, FFRDCs, university-affiliated research centers (UARCs), and universities (DoD officials, personal communication).

Question 2: Potential to Increase the Use of Analytic Capabilities to Improve Acquisition Outcomes

Q2. What is the potential to increase the use of analytic capabilities to improve acquisition outcomes?

Findings: Options include:

- Develop a data-analytics strategy between acquisition domains
- Continue maturing data collection, access, and analytic layers on systems
- Expand analytic tool availability
- Create statutory authorities for external analyst protected access to data
- Establish policy of full data access for analysts
- Improved decisionmaker incentives and understanding of data analytics may improve use of available data
- Expand types of analysis

Perspectives:

- Commercial advanced data analytics do not always apply to government
  - Even industry has problems implementing data analytics
- Need more predictive indicators
- Data analytics will not eliminate acquisition risks

Question 2 asks about "the potential to increase the use of analytical capabilities for acquisition programs and offices to improve acquisition outcomes" (U.S. House of Representatives, 2016, p. 1125).

Opportunities to Increase Use of Analytics

Based on some known challenges, expertise, and new ideas, we identified a number of options that the DoD could consider for exploiting data analytics to improve acquisition.

Although various acquisition domains have data *management* strategies (including plans for improving analytic layers on data systems), no overarching acquisition *data analytics* strategy provides key strategic questions *across* acquisition domains that need to be answered, along with the underlying data that must be collected to enable the analysis. Such a strategy could serve to inform investments while seeking to ensure that cross-domain linkages are addressed and coordinated. Also, some of the data management strategies lack explicit consideration of what key strategic questions need to be answered, thus informing what analyses, analytic capabilities, and data are needed. These strategic plans could also be improved by including consideration of the following elements.

Generally, the DoD has made progress on maturing its data collection and access while adding analytic layers to existing systems (see the prior slide). These activities tend to leverage private-sector best practices (except for some cultural data-sharing issues). Data governance is recognized as being important, and the DoD has made significant progress in identifying authoritative sources, standardizing data definitions across the department, and identifying key data uses (especially statutory requirements) to inform data selection and management. Maturity varies, and more could be done, as evidenced by various plans across the DoD. For example, although OSD's and the Air Force's programmatic business information systems are more mature, the Army's plan to evolve toward a system with open data interfaces and an analytic layer is being prototyped with final implementation decisions and funding pending. Also, the DoD's program information managers recognize that a more comprehensive understanding of the needs (use cases) to drive data collection and analysis is needed, especially given recent acquisition reforms and organizational changes. These practices have created an environment more conducive to analytics; however, some obstacles still impede DoD's continuing improvement in acquisition analytics. For example, DoD acquisition data management continues to be stovepiped by components and functional communities. Few formal (i.e., established) mechanisms exist to deal with cross-functional issues. This stovepiping has also resulted in varying levels of maturity in data management practices across the acquisition community, making it even more difficult to accomplish such tasks as sharing between major information systems.

Expanded availability of the latest analytic software tools would also help increase analytic capabilities. DoD cybersecurity concerns limit the installation of the wide and expanding range of commercial software. One approach is to increase the testing and create a DoD-wide list of approved analytic tools that can be installed on DoD computers. Alternatively, the use of virtual computing environments in which COTS software can be run in isolation from DoD networks could be expanded. Virtual environments solve the problem of isolating security concerns, but that does impede data and information flowing out of the virtual environments.

Access to data for both internal government analysts and nongovernment analysts needs continued and significant improvement. Progress is mixed. Some progress has been made in establishing authoritative data sources with open data-exchange interfaces internal to the DoD, but this progress is often limited to statutorily required programmatic data on major programs and data destined for public release (e.g., contracting awards in the Federal Procurement Data System–Next Generation [FPDS-NG]). Other data (e.g., programmatic, contracting, and engineering data on smaller programs and engineering details on larger programs) are not widely shared between military departments or with OSD. Also, obtaining user accounts on individual systems is laborious, because it must be done one account at a time, and obtaining accounts on systems outside one's organization can be difficult or essentially impossible. Barriers are engrained in a culture driven by concerns that include security (both directly and from data aggregation), excessive oversight, interventions by and distractions from staff outside the immediate organization, and loss of limited management and execution time from exposure and attention. These barriers apply recursively up the acquisition chain of command and to Congress (the program manager might be busy solving problems and would prefer to focus on management by limiting time spent on reporting to the program executive officer [PEO]; this same concern applies from the PEO to the service acquisition executive [SAE], from the SAE to OSD, and from the SAE and OSD to Congress). All parties recognize some needed level of oversight and thus data sharing, but the belief is that current levels impose burdens on time and resources that are better spent in direct management.[1] In essence, these cultural barriers to data sharing attempt to introduce program efficiency by limiting data sharing and access.

In contrast, the private-sector best practices (see the discussion below, on Congress's question 4) that are being pursued strategically by the DoD's information system managers have reduced burden through more-efficient data extraction directly from operating functions, standardized data definitions, designation of the authoritative information source for each data element (to avoid arguments or misleading analysis or decisions), open interfaces to share authoritative data efficiently and to minimize data entry to once (or less, if automatically pulled), selection of which data to make available in each system based on key decisional needs, and provision of integrated software analytic and visualization tools for general-purpose data analytics. This approach requires continued strategic support from leadership, continued investments, and more time to fully implement. It also needs stronger policies that clarify who "owns" the DoD's acquisition data and thus who can have access and for what reasons (e.g., whether acquisition data are *common enterprise data* and how to reduce excessive intervention on programs). How to successfully address all these concerns requires further policy analysis, but designation as *common enterprise data* could be a first step, given that 10 U.S.C. 2222(e) specifies that such data are available across the DoD.

For external analysts, one partial solution would be for Congress to make permanent the access for DoD FFRDCs to sensitive data (e.g., proprietary and other controlled unclassified information, as reflected in Section 235 of the FY 2017 NDAA [Pub. L. 114-328, 2016]).[2] This is only a partial solution, since other approaches are needed to address access for nongovernment employees of UARCs, labs, test centers, and support contractors. We make further observations below in response to Congress's questions related to data anonymization and more-automated data collection and access.

A related idea is to develop DoD-wide categories for data access. Currently, access must be adjudicated and granted information system by information system. Ideally, individuals (either

---

[1] Sources include discussions with DoD officials and the experience of the authors.

[2] For full disclosure, the RAND Corporation operates three DoD FFRDCs, so the quality of its research could benefit from such access. Per Section 235 (Pub. L. 114-328, 2016):

> (g) Sensitive Information Defined.—In this section, the term "sensitive information" means confidential commercial, financial, or proprietary information, technical data, contract performance, contract performance evaluation, management, and administration data, or other privileged information owned by other contractors of the Department of Defense that is exempt from public disclosure under section 552(b)(4) of title 5, United States Code, or which would otherwise be prohibited from disclosure under section 1832 or 1905 of title 18, United States Code.

government employees or contractors) within proper categories working to analyze data across the DoD could be designated by an approval authority, then automatically granted access to various data systems (e.g., through a shared access certification and perhaps even single sign-on).

Furthermore, improved decisionmaker incentives and understanding of data analytics may improve the use of available data in decisions. We recognize that decisionmakers must balance risks, equities, and other strategic factors when making decisions (i.e., decisionmaking is not simply the blind application of process), but sometimes other priorities take precedence, or the data might not be fully appreciated. Continued grassroots training of career managers and rising leaders about the strengths and limitations of data analytics will help. How to improve political appointees' understanding of the strengths and limitations of data analytics is more challenging because they are often in office for only short periods. An approach for political appointees might be to explicitly include an overview of the DoD's acquisition data analytics capabilities and their strengths and weaknesses in orientation materials, as well as in briefings and memoranda (i.e., do not assume familiarity, and be more explicit about strengths, uncertainties, and limitations of the data analysis being presented). All leadership categories may benefit from strategically ensuring that they have the incentives and authorities that enable them to most appropriately balance insights from data analytics against other strategic considerations (e.g., political, budgetary, mission, urgency, and threats).

Finally, expanding the type of data analysis topics has the potential to improve acquisition outcomes. Although the topics depend on current leadership priorities and needs, the following examples illustrate some possible areas that could be further explored.

- **Externalities:** Some data analytics results do not delineate the effects of external factors from those internal to the acquisition system. For example, operation and sustainment cost growth can result from poor fuel efficiency (a technical and design factor within the acquisition community's area of responsibility) or from increases in fuel cost (external to the DoD, from the global marketplace). Improved data analytics capabilities and approaches could be developed to try to separate these factors to inform decisionmakers on what is driving cost growth and whether the acquisition community is responsible or not.
- **Mission-level (versus program-level) portfolio analysis:** Outcome-based portfolio analysis should be performed (e.g., to understand technical, budgetary, and schedule implications to mission or kill-chain effects from cross-program dependencies). The DoD is exploring how to shift from assessing acquisition program performance in isolation to assessing performance as it pertains to the integrated set of programs that field mission-level capabilities (e.g., kill chains). Such analysis offers the potential to view acquisition from the perspective of integrated warfighter capability effects and outcomes rather than delivery of individual systems. Integrated analysis could help decisionmakers better understand dependencies between systems so that technical, budgetary, and schedule decisions can be made to maximize the integrated warfighting capability. For example, such a shift could enable better budgetary, schedule, and quantity decisions based on how a smaller program fits within an overall kill chain. Otherwise, smaller programs may

receive insufficient support because leadership and Congress do not have the full picture of the importance of that program in larger warfighter capabilities.

- **Framing assumptions:** Framing assumptions analysis and associated metrics tracking should be fully implemented (Arena et al., 2013; Arena and Mayer, 2014; Department of Defense Instruction 5000.02, 2017). The DoD and its FFRDCs have developed the concept of assessing the key conceptual risks of a program so that leadership can understand the risk (i.e., the "big bets") they are taking when approving a program. Although codified in current DoD policy (Department of Defense Instruction 5000.02, 2017), the use of framing could be expanded to allow Congress to better understand the risks involved in a new program it authorizes and funds. Also, it is unclear whether the current trends away from approaches such as in Department of Defense Instruction 5000.02 (e.g., in the "Middle Tier" of acquisition, noted in 10 U.S.C. 2302) need to be clarified so that the utility of understanding the framing assumptions in a program at initiation are assessed and articulated to decisionmakers. Finally, tracking these risks could be made more explicit so that early indicators of fundamental conceptual risks are detected as soon as possible.
- **Performance analysis:** The DoD could continue making inroads using data analytics to understand institutional performance trends and what may be driving them. Such quantitative and qualitative analyses seek to understand statistically significant trends rather than focusing on the performance of an individual case or program. For example, the DoD published four reports of such analysis a few years ago (see Under Secretary of Defense for Acquisition, Technology, and Logistics, 2013, 2014, 2015, 2016a) and is rebuilding a capability to conduct such analysis in the Office of the Under Secretary of Defense for Acquisition and Sustainment.
- **Data needed:** Data collection and management involve significant effort and investments, so analysis to better understand which data are needed can inform acquisition efficiency efforts and ensure that key data to inform decisions and data analytics are available. Such efforts start with the important questions and decisions that need to be informed, but analysis includes assessing the relative costs and benefits of obtaining data through alternative mechanisms.

Perspectives

One of the drivers for Congress's question appears to be the recent commercial advances and successes in data analytics, including advanced analytics, such as artificial intelligence, machine learning, and big-data techniques. We found, however, that simplistic applications of many commercially successful techniques do not easily apply to data analytics for acquisition program management. The data and domains differ. For example, the extremely large evaluated data sets needed to train neural networks do not exist in the DoD, and acquisition programs tend to fail for different reasons, so the challenge is not one of categorization to identify problems. There are, however, applications where commercial advanced analytics may apply (e.g., in text understanding and network analysis of supply chains), and the DoD is conducting research to explore those possibilities.

Also, the suggestion of pursuing a data analytics strategy is based on insights from industry and the data analytics consulting field. Cases of (sometimes spectacular) commercial successes from data analytics often come from a combination of well-thought-out analytic planning by leadership, a clear definition of *value* from implementing analytics for the organization, an adequately trained workforce (both data scientists and functional business experts), well-defined use cases, and appropriate tool selection. Mismatches and a lack of strategic planning and support lead to lesser-known but not infrequent failures.

A different challenge is the general lack (and need) for more predictive (leading) indicators of acquisition problems. Most available indicators used by the DoD are descriptive (lagging), although there are some examples of leading indicators, such as contractor risk indicators, earned value,[3] and cross-program performance indicators (e.g., extrapolating known concerns to similar acquisition activities). There are no easy candidates for new leading indicators, so this is a research challenge, not an immediate opportunity.

Although there are opportunities to expand data analytics to improve acquisition outcomes, it is important to remember that defense acquisition inherently involve risks. Data analytics can improve insight into those risks, but they cannot be eliminated by increased analytic capabilities. Decisionmakers are judging and taking risks to field new, advanced capabilities. Other strategic factors influence decisions besides risk elimination (e.g., advancing the state of the art, driving faster delivery when lives are at risk, keeping pace or staying ahead of evolving threats, and trying to pursue new capabilities despite limited budgetary and workforce resources).

---

[3] Note that earned value is both a leading and lagging indicator. It is leading in that it can be used to inform estimations of completion time and cost, but the actual performance reports are lagging (see, for example, Ferguson, 2008).

Question 3: R&D Funding for Analytic Capabilities for Acquisition

Q3. How much R&D funding is there to develop and implement data analytic capabilities?

Findings:

- FY 2019 DoD RDT&E budget request:
  - Estimate: roughly $200 million (+/- 80%) across 31 program elements (PEs)
    - *Found specific discussion related to data analytics for acquisition*
  - $520 million/year for IT systems (up from $313 million in FY 2017)
    - *Provides data and sometimes have analytic layers*
- Some SBIR and STTR topics in January 2019 solicitations for acquisition data analytics
- Found anecdotes of exploratory research on advanced analytics for acquisition

Notes:

- This does not include R&D outside acquisition proper
  - Military operations, budgeting, requirements, or intelligence analysis in support of acquisition
- Again, direct application of advanced commercial data analytics is challenging

NOTE: SBIR = Small-Business Innovation Research; STTR = Small-Business Technology Transfer.

Question 3 asks about "the amount of funding for intramural and extramural research and development activities to develop and implement data analytics capabilities in support of improved acquisition outcomes" (U.S. House of Representatives, 2016, p. 1125).

Findings

FY 2019 DoD Budget Request for Research, Development, Test, and Evaluation

The DoD's chart of accounts for RDT&E does not specifically track R&D for data analytics. However, we were able to analyze available information in the FY 2019 budget request to get a rough estimate of DoD-wide investments. Although still approximate, these estimates give a measure across the DoD rather than requesting anecdotal examples from DoD components.

One part of our analysis examined the in-depth RDT&E justification books for the DoD's FY 2019 budget request. Using keyword searches, we identified 31 program elements (PEs) that mentioned data analytics targeted specifically for defense acquisition. Using our expertise, we read each of those PEs to estimate how much of the PE is focused on developing and implementing data analytics (i.e., most, some, or a little). Given those rough assessments, the total investments may be about $200 million for FY 2019 (plus or minus roughly 80 percent); however, that is a rough-order-of-magnitude approximation, given the fidelity of the qualitative descriptions. Still, this gives an indication that there are activities across the DoD with a rough sense of the magnitude of R&D funding.

Another piece of our analysis focused on investments in major IT systems (not desktop computing), discussed earlier in response to Congress's first question. Those budget request details indicate both the topic areas (acquisition or logistics and supply-chain management) and the funding type (RDT&E, in this case). The request for FY 2019 and FY 2020 were $520 million and $524 million, respectively (in then-year unadjusted dollars). Those investments are an increase from the $313 million and $411 million reported for FY 2017 and FY 2018.

SBIR and STTR Solicitations in January 2019 Solicitations

We also found four related topics in the January 2019 SBIR and STTR solicitations for design tools and production improvements and within acquired systems. We note that SBIR and STTR solicitations might not result in awards, and any such initial awards are relatively small compared with PEs, but we include them to try to measure DoD interest in R&D for acquisition data analytics.

Exploratory Research on Advanced Analytics

In addition, we found anecdotes of DoD research efforts that explore advanced analytic applications in acquisition, but these are smaller exploratory efforts below the size of PE budgets. Evidence for the success of these efforts is pending, so it is hard to predict their value or how large these investments may become.

Note that these investments do not include R&D for military operations or other areas outside acquisition (e.g., budgeting, requirements, or intelligence).

Question 4: Private-Sector Best Practices for Minimizing the Collection and Delivery of Data

Q4. What are private-sector best practices that could minimize collection and delivery of data by, from, and to government organizations?

Findings:

- Private-Sector best practices include:
  - Identify specific questions that leadership needs to answer in order to make informed decisions
  - Plan and prioritize what data are needed using an organization's data strategy
  - Define data and establish common definitions between organizations using that data
  - Designate single authoritative source for each datum or dataset, then share existing authoritative datasets between systems via technical means
  - Emphasize that data are corporate-wide assets, not owned by local units

DoD adoption to date:

- DoD is pursuing most of these practices
  - Common PM tool suites are less common but being pursued
  - Other acquisition IT systems provide data feeds (e.g., contracting, auditing)
  - Open systems and data sharing have reduced duplication and ensured common data
  - Data-element sensitivity meta-labeling
- Data sharing and ownership challenges remain

Opportunities:

- Continue pursuing relevant best practices from the private sector
- Clarify that data are DoD-wide assets and address disincentives to data sharing
  - Clarify oversight extent and roles
  - Continue research on sensitivity upgrades from data-aggregation

NOTE: PM = program management.

Question 4 asks about the "potential improvements, based on private-sector best practices, in the efficiency of current data collection and analysis processes that could minimize collection and delivery of data by, from, and to government organizations" (U.S. House of Representatives, 2016, p. 1125).

Findings

A number of private-sector best practices (identified through literature reviews, prior research, and attendance at domain meetings) could improve DoD efficiency by minimizing the collection and delivery of data.[4]

One practice is to objectively plan and prioritize what data are essential to collect in the first place (i.e., define the organization's data strategy) by identifying specific questions that leadership needs answered to make informed decisions, the essential data options needed for analysis, and the costs and benefits of these alternatives (Nott, 2015; *MIT Sloan Management Review*, 2016; vander Meulen and McCall, 2018; DalleMule and Davenport, 2017). One specific

[4] These are best practices in that they are from consulting companies that assess, survey, and review the field for lessons learned and common practices. We also found them to be relatively consistent.

approach is to identify and prioritize such use cases as vehicles for planning and management (Fleming et al., 2018).

A second set of practices is defining the data, establishing common definitions between organizations using those data, and automatically collecting data from operational systems for subsequent analysis and use by management and oversight (Taylor, 2016). Automatically collecting data alleviates manual data reporting and can provide more-accurate, more-current, and more-detailed data on activities and functions. Transmittal of such data is performed seamlessly by interconnected major IT systems.

A third practice is to designate which data system will be the single authoritative source for a piece of information (DalleMule and Davenport, 2017), then share that information via technical means to other systems that need it. This increases transparency and ensures that everyone is using the same data while reducing duplicative, and potentially erroneous, data entry.

Finally (and perhaps most important), the private sector emphasizes that data are corporate-wide assets. They are not owned and controlled by local units but are generally available widely (with appropriate privacy protections) to improve the efficiency of the organization (Svensson, 2013).

We assessed the DoD against these practices by comparing the DoD's status (as identified in prior research [Riposo et al., 2015; McKernan et al., 2016, 2017, 2018] and recent DoD information-management meetings against these best practices. Comparison is straightforward, identifying whether key private-sector practices are being used by the DoD.

DoD Adoption to Date

The data managers across the DoD are pursuing many of these practices; however, the level of maturity of these practices varies widely across the DoD's acquisition community. Use cases are generally used by information managers to identify and manage key data. Designating authoritative sources and sharing data between systems has become the standard approach by information managers, and its use is maturing in many systems across acquisition functions. Common PM software suites that can automatically share such data as program cost, schedule, and status information are less common but are reflected in current DoD plans. Considerations, such as ensuring that draft data are not transmitted until they are finalized, will have to be made when selecting or designing software suites.

Although the DoD has made progress in opening and sharing its acquisition data systems, various challenges to sharing remain. The general culture against sharing is driven by various disincentives and concerns, including security (e.g., sensitivity elevation from data aggregation), trust (e.g., that oversight support staff will use data to micromanage projects, personnel will provide to unauthorized users), and data labeling (e.g., a general lack of metadata on the sensitivity level of individual data elements).[5] Another major concern is that information that was not cleared for public release will end up in the public domain.

Opportunities

Opportunities for the DoD include continuing to pursue these best practices, addressing the disincentives to share, adding incentives to share, and developing approaches to resolve security and sensitivity issues.

The DoD could also address the disincentives to data sharing by emphasizing that data are DoD-wide assets—for example, by further clarifying oversight extent and roles by acquisition level (I–IV) so that smaller programs are clearly delegated to lower levels in the acquisition chain of command and elevated only when problems occur and when those problems clearly require higher-level help. Also, the DoD could continue research on how to understand data-sensitivity upgrades when data are aggregated (as when made available for analysis).

Question 5: Steps Being Taken to Expose Anonymized Data to Researchers and Analysts

Q5. What steps are being taken to expose anonymized data to researchers and analysts?

Findings:

- Generally, DoD is not anonymizing acquisition data for various reasons
  - A counter-example is Personally Identifiable Information (PII)
- DoD is:
  - Identifying available data
  - Improving internal data transparency through technical means
  - Adding researcher and analyst accounts on data systems
  - Removing classified and sensitive data elements before publishing

Perspectives:

- Anonymization has significant limitations
  - Some anonymization can be broken with modern correlation tools
  - Meta-data are required for correlation analysis and cannot be anonymized
  - Data sensitivity is often not stored at the data element level
  - Procedures for categorizing and handling sensitive data are complicated, slow, and not well understood; incentives impede progress
  - Aggregating data can increase data sensitivity and classification

Opportunities:

- Process data in a clean room
  - Limits use of anonymization-breaking tools
- Possible option: blind analysis
  - Algorithms run in a black box, only showing the results
  - Challenges include data cleaning and understanding
  - Research would be needed
- *(See prior slide for other data sharing opportunities)*

[5] Sources include meetings with DoD officials, prior research (Riposo et al., 2015; McKernan et al., 2016, 2017, 2018), and the experience of the authors.

Question 5 asks about the steps "being taken to appropriately expose acquisition data in an anonymized fashion to researchers and analysts" (U.S. House of Representatives, 2016, p. 1125).[6]

Findings

Although the DoD has made some progress in improving data sharing, it is not generally anonymizing data, for various reasons discussed below. One major counterexample is personnel data (including AWF data), which are often sensitive, PII with legal releasability restrictions. Personnel data are often anonymized, but even such data are hard to get directly from the source (although some anonymized data can be obtained indirectly through Office of Personnel Management sources).

Although not widely using anonymization per se as a sharing technique, the DoD is increasing data availability by identifying what data are available, improving access through technical means and interfaces, adding the ability for some researchers and analysts to obtain accounts on some data systems, and transmitting unrestricted data after extracting sensitive and classified data (as is done, e.g., with contracting data information sent to FPDS-NG).

Perspectives

There are practical reasons why anonymization has not been widespread. Anonymization is not always reliable; advances in analytic tools can sometimes identify data (e.g., in personnel data). Also, much of the metadata that would be removed in anonymization is important for trying to identify potential causes of identified trends. In addition, DoD data generally lack data-sensitivity metadata at the data-element level, making it hard to select what data cannot be shared and why. Furthermore, government procedures for categorizing and handling sensitive data are complicated, slow, and not well understood by staff, and incentives drive conservatism to block sharing (e.g., what exactly can and cannot be asserted as proprietary information by a contractor, how can markings be changed, and what are the personal risks involved?).

Opportunities

One approach to limiting the use of tools that could break the anonymization is to require researchers to come to a clean room (controlled environment) in which the tools they employ are monitored and only the results of the analysis (not the anonymized data) can be removed. This is practiced in some DoD cases for anonymized personnel data, although other anonymized data are available publicly (e.g., see U.S. Office of Personnel Management, undated-b).

---

[6] Some kind of data protection is often needed, since acquisition data can be source-selection sensitive, classified, operationally sensitive, personally identifiable information (PII), predecisional, trade secrets, proprietary to contractors or commercial companies, restricted to the Committee on Foreign Investment in the United States, or other types of sensitivities.

One possible alternative to straightforward anonymization might be to allow researchers and analysts to develop algorithms that run on the raw data in a protected "black box" environment, exposing the results only to the researchers. This is just a notion at this point, since issues of data cleaning, data understanding, and ensuring that the resulting analytic output do not divulge sensitive information would need to be developed.

Question 6: DAU on Data Analytics and Other Evaluation-Related Methods

Q6. Do training institutions include appropriate courses on data analytics and other methods and their application to defense acquisitions?

Findings:

- Yes, DAU, NPS, NDU, AFIT, and partner universities and institutions* offer:
  - Applied methods and tools courses
  - Applied data analytics courses
  - Generic data analytics courses and electives
  - Advanced analytics courses are predominantly at NPS and partner institutions
- Enrollments at DAU in FY 2018 indicates a reasonable stratification:
  - *~150,000: Applied methods and tools courses*
  - *~ 60,000: Applied data analytics courses*
  - *~ 3,300: Generic data analytics courses*
- These courses should help staff understand how to request analysis and understand the results

* Georgia Tech, American University, George Mason University, Georgetown University, George Washington University, Johns Hopkins University, Stanford, University of Michigan, Google, IBM, and DoD Cyber Crime Center (DC3) Cyber Training Academy

Perspectives:

- Not everyone in acquisition can or should become a data scientist
- On-the-job training is also important
- Analytic layers on information systems are fairly intuitive and come with online help and training sessions
- As in industry, few people understand both data science and the application area
- Did not assess quality of courses

Opportunities:

- Implement rotations in analytic-based offices
- Better inform personnel of available analytics-based courses
- Encourage analytics-based training

NOTE: NPS = Naval Postgraduate School; NDU = National Defense University; AFIT = Air Force Institute of Technology.

Question 6 asks "whether the curriculum at the National Defense University, the Defense Acquisition University, and appropriate private-sector academic institutions includes appropriate courses on data analytics and other evaluation-related methods and their application to defense acquisitions" (U.S. House of Representatives, 2016, p. 1125). We reviewed the curricula at four defense institutions: NDU, DAU, NPS, and AFIT. We also examined acquisition-related offerings at five external universities: the Georgia Institute of Technology (Georgia Tech), American University, George Mason University, Georgetown University, and the George Washington University. Finally, we reviewed courses with the following strategic partners that DAU established: the Johns Hopkins University, Stanford University, University of Michigan, Google, IBM, and the DoD Cyber Crime Center (DC3) Cyber Training Academy.

Findings

DAU, NPS, and AFIT are the primary acquisition training institutions for the DoD (whereas NDU is primarily a defense strategy university). These institutions offer a variety of acquisition courses with analysis content that ranges from theory and processes to the addition of methods, tools, and in-depth applied data analytics. For example, a course in cost analysis is categorized as in-depth applied data analytics. To give a sense of the relative scope, these institutions offer courses in these four categories:

- acquisition theory and processes only: 17 percent (99 courses total out of 578 examined) among the four institutions
- acquisition theory, processes, and methods: 18 percent (103 courses total) among the four institutions
- acquisition processes and methods with tools: 12 percent (70 courses total) among the four institutions
- data analytics and other evaluation-related methods applied to acquisition: 24 percent (139 courses total) among the four institutions.

They also offer 25 general-purpose data analytics courses (not directly targeted at acquisition), constituting 4 percent of the total courses reviewed. Note that the 578 courses we assessed are restricted to areas of interest that we assessed; thus, for example, we did not include the entire curriculum outside acquisition at NPS.

Also, DAU partners with other universities and institutions to offer additional courses, including advanced analytics. Partners include Georgia Tech, American University, George Mason University, Georgetown University, George Washington University, Johns Hopkins University, Stanford University, University of Michigan, Google, IBM, and the DC3 Cyber Training Academy.

Attendance numbers at DAU also convey a sense of engagement: Enrollments in applied data analytics courses in FY 2018 were about 60,000, with about 209,000 enrollments in the other three applied categories total. General analytic course enrollments were about 3,300 in FY 2018, so there is some penetration in general data analytics outside direct application.

The offerings at the five external universities ranged from applied acquisition training to courses related to public policy. Finally, DAU's education partners offer in-depth general data analytics courses that quickly leverage partner offerings without having to create duplicative capabilities at DAU. These courses range from statistical programming to machine learning, text mining, serverless data analysis, and cyber training.

Perspectives

Although these institutions offer a range of data analytics training, it is important to keep training in perspective. Not everyone in acquisition can (or should) become a deep data scientist. These applied and general-purpose courses should increase the ability of the AWF to conduct simple analysis while becoming smart consumers of analysis conducted by specialists.

Note also that formal institutional training is not the only mechanism for improving the data analytics capabilities of the AWF. On-the-job training and mentoring are also important. Also, analytic tools, such as the analytic layers being added to acquisition information systems, are fairly intuitive and come with online help and training sessions.

To be successful in applying advanced data analytics and other methods to acquisition requires both data analytics expertise and domain knowledge in acquisition. People with these skills are in short supply, especially in the government AWF. Although this may seem to be a serious problem, industry faces the same type of shortfall. A more pragmatic expectation for training may be a gradation of analytic skills within the AWF, ranging from a few deep analysts, to many people being able to conduct simple analyses with easy-to-use tools (e.g., in data system analytic layers), to most people having at least a rudimentary ability to intelligently consume the results of analysis.

Opportunities

Potential opportunities and actions (by stakeholder)

Congress

- Declare in 10 USC 2222(e) that acquisition data are "Common Enterprise Data" and thus must be shared across DoD
- Make permanent the FFRDC access to sensitive data under Sec. 235 of the FY 2017 NDAA

USD(A&S) with CMO, CIO, CDOs, and SAEs

- Address disincentives to data sharing
- Enable DoD-wide access for appropriate analysts to sensitive data
- Facilitate access to analytic tools through virtual computing environments and an embellished list of standard software approved for installation on DoD computers
- Continue R&D on improved data and analytic systems and new applications in acquisition
- Develop a data-analytics strategy across acquisition domains

DoD information managers

- Continue pursuing PM and process suites with data outputs to feed oversight information systems
- Continue maturing data collection, access, and analytic layers on data systems
- Continue compiling and sharing catalog of available data

Defense acquisition training institutions

- Continue offering data science courses and applied data analytics for staff, management, and rising leaders

NOTE: CMO = chief management officer; CIO = chief information officer; CDO = chief data officer.

Finally, we grouped our ideas of potential opportunities and actions identified in our study according to what stakeholder group would implement them.

Congress

Two opportunities generally apply for congressional consideration. First, Congress could clarify in 10 U.S.C. 2222(e) that all acquisition and sustainment data are common enterprise data and thus available across the DoD. This would provide additional statutory authority for access and may improve sharing. Second, Congress could make permanent the three-year pilot established by Section 235 of the FY 2017 NDAA that grants analysts in FFRDCs access to sensitive data.[7] This pilot provides an efficient mechanism to establish access to, and protection of, sensitive data beyond government employees.

DoD Leadership

Members of DoD leadership (Under Secretary of Defense for Acquisition and Sustainment, with the CMO, CIOs, CDOs, and SAEs) have several options they could consider. First, they could address disincentives for data sharing, including clarifying oversight extent and roles to minimize concerns of micromanagement above the PEO, expanding security metalabeling at the data-element level (rather than for a whole corpus of data), and further addressing security concerns by clarifying sensitivity upgrades from data aggregation. Second, they could direct full access to acquisition data across all DoD systems for analysts who are assessing broader institutional trends and performance correlates; this is especially true for lower-sensitivity data (e.g., predecisional information). Third, analytic tool access could be facilitated through wider use of virtual computing environments for analysis and an embellished list of standard software approved for installation on DoD computers. Fourth, R&D could be continued on both improved data analytics systems and new applications. Finally, strategy could be expanded beyond individual functional domains to develop a cross-domain data analytics strategy to integrate these kinds of improvements.

DoD Information Managers

For the DoD's information managers, the opportunities appear to surround continuation of the efforts to date and expansion to all core acquisition functional areas. These include (with leadership support and funding) continuing to pursue PM and a process suite that could automatically collect and distribute relevant data; continuing to mature data collection, access, and analytic layers on data systems; and continuing to compile and share information about where data reside and how to access them.

Defense Acquisition Training Institutions

Similarly, the defense acquisition training institutions could continue offering up-to-date applied and general data analytics courses, both as part of the core training curriculum and as electives to improve the DoD's organic AWF capabilities.

---

[7] For full disclosure, we work for an FFRDC, so the quality of our research would benefit from such access.

Steps for Further Research

NOTE: SETA = systems engineering and technical assistance.

Although the ideas outlined on the prior slide provide some specificity into potential opportunities for improving data analytics in the DoD, some of these and other ideas require further research to develop actionable recommendations for implementation.

Congress

For Congress, we note in our briefing that sometimes other considerations led to acquisition decisions that ran counter to available certain data analytics results, apparently because of conflicting incentives and other strategic motivations. Research could further clarify what drives these decisions and how Congress could better balance decision incentives and authorities. For example, what is the right balance between streamlined oversight and independent checks and balances? Also, a broader need exists beyond FFRDCs for efficient access to (and protection mechanisms for) other sensitive data for analysis. Research is needed to develop options for UARCs, contractors working for defense laboratories, and SETA support contractors.

DoD Leadership

Further research could also inform actions by DoD leadership—for example:

- methods to address disincentives to data sharing

- policy and process analysis on data aggregation and classification upgrades, including aggregation of sensitive, unclassified information to ensure more-consistent application across the DoD
- policies and approaches for granting DoD-wide access to different DoD information systems for government and contractor analysts
- identification of the minimum data needed, at what level, and for what purposes
- detailed analysis to create a cross-domain DoD data-analytics strategy.

Acquisition Training Institutions

For acquisition training institutions, it would be useful to better understand not only what types of data analytics courses are offered but also their quality and practical utility, especially given the diverse nature of the AWF and its range of technical backgrounds and skills.

DoD Analysts

The five types of new or expanded analyses presented earlier could improve acquisition management and decisions.

Acknowledgments

Acknowledgments

- OSD sponsors and staff
  - Hon. Kevin Fahey, David Cadman, Brian Joseph, Mark Krzysko, Phil Rodgers, Nicolle Yoder, Rhonda Edwards, Mark Hogenmiller
- DAU/NDU
  - James Woolsey, Leslie Deneault, Richard Hoeferkamp, Bill Parker, David Pearson, Mary Redshaw
- Department of the Army
  - Ray Gagne, MAJ Thomas P. Dirienzo, Michael J. Gully, Michael A. Larsen, Alpesh Patel, COL Joseph W. Roberts, Keith Samuels, COL Gregory Smith, Leo Smith
- Department of the Navy
  - Barbara (BJ) White-Olson, William Bray, CDR William Benham, Robert Borka, Charlotte Helan, Dale Moore, Linda Wack
- Department of the Air Force
  - John Miller, Mildred Bonilla-Lucia, Ralph DiCicco, and Col. David J. Sanford
- RAND
  - Frank Camm, Scott Comes, Justin Grana, Kristin Lynch, Lauren Mayer, Igor Mikolic-Torreira, Brian Persons, Joel Predd, William Shelton, Jia Xu, Sarah Meadows, and Obaid Younossi

We thank our project sponsors, David Cadman and Brian Joseph, Office of the Under Secretary of Defense for Acquisition and Sustainment (OUSD[A&S]), for their support throughout this effort. We also thank Kevin Fahey, Mark Krzysko, Phil Rodgers, Nicolle Yoder, Rhonda Edwards, Mark Hogenmiller in the OUSD(A&S); James Woolsey, Leslie Deneault, Richard Hoeferkamp, Bill Parker, David Pearson, and Mary Redshaw, from the Defense Acquisition University and National Defense University; Ray Gagne, MAJ Thomas P. Dirienzo, Michael J. Gully, Michael A. Larsen, Alpesh Patel, COL Joseph W. Roberts, Keith Samuels, COL Gregory Smith, and Leo Smith, in the Department of the Army; Barbara (BJ) White-Olson, William Bray, CDR William Benham, Robert Borka, Charlotte Helan, Dale Moore, and Linda Wack, in the Department of the Navy; and John Miller, Mildred Bonilla-Lucia, Ralph DiCicco, and Col. David J. Sanford in the Department of the Air Force.

We appreciate the support from a number of our colleagues at RAND, including Frank Camm, Scott Comes, Justin Grana, Kristin Lynch, Lauren Mayer, Igor Mikolic-Torreira, Brian Persons, Joel Predd, Jia Xu, Sarah Meadows, and Obaid Younossi. Scott Comes, Justin Grana, Christopher Mouton, and William Shelton provided very helpful peer reviews. Rebecca Fowler provided very useful edits. All errors, however, remain those of the authors.

Abbreviations

A&S	Acquisition and Sustainment
AAG	Advanced Arresting Gear
AF	Air Force
AFIT	Air Force Institute of Technology
AIR	Acquisition Information Repository
AoA	analysis of alternatives
API	application programming interface
ASA(FMC)	Assistant Secretary of the Army for Financial Management and Comptroller
ASAF(FMC)	Assistant Secretary of the Air Force for Financial Management and Comptroller
ASN(FMC)	Assistant Secretary of the Navy for Financial Management and Comptroller
AU	American University
AWF	acquisition workforce
BI	business intelligence
CAD	computer-aided design
CADE	Cost Assessment Data Enterprise
CAE	Component Acquisition Executive
CAPE	Cost Assessment and Program Evaluation
CDO	chief data officer
CE	cost estimating
CIO	chief information officer
CMO	chief management officer
CONEMPS	concepts of employment
CONOPS	concept of operations
COTS	commercial off-the-shelf
CPARS	Contractor Performance Assessment Reporting System

CTR	contractor (support)
D&A	data and analytics
DAE	defense acquisition executive
DAMA-DMBOK	Data Management Association International's Data Management Body of Knowledge
DAMIR	Defense Acquisition Management Information Retrieval
DARPA	Defense Advanced Research Projects Agency
DAU	Defense Acquisition University
DAVE	Defense Acquisition Visibility Environment
DC3	Department of Defense Cyber Crime Center
DCMA	Defense Contract Management Agency
DIBNow	Defense Industrial Base Now
DMM	Data Maturity Model
DoD	U.S. Department of Defense
DoDI	Department of Defense Instruction
DRDW	DoD Resources Data Warehouse
ECSS	Expeditionary Combat Support System
Engr.	engineer
ERP	enterprise resource planning
eTools	Electronic Tools (Defense Contract Management Agency)
EVM-CR	Earned-Value Management—Central Repository
FCS	Future Combat System
FFRDC	federally funded research and development center
FM	financial management
FPDS-NG	Federal Procurement Data System—Next Generation
FY	fiscal year
GCSS-MC	Global Combat Support Systems—Marine Corps
GMU	George Mason University

IDECM B2/B3	Integrated Defensive Electronic Countermeasures, Block 2/3
IT	information technology
JLTV	Joint Light Tactical Vehicle
KM/DS	Knowledge Management/Decision Support
LCS	Littoral Combat Ship
LIMS-EV	Logistics Installation Mission Support—Enterprise View
M&S	modeling and simulation
MCSC	Marine Corps Systems Command
MDA	Milestone Decision Authority
MDAP	Major Defense Acquisition Program
MIL	military (personnel)
MIT	Massachusetts Institute of Technology
MS	milestone
NAVAIR	Naval Air Systems Command
NDAA	National Defense Authorization Act
NDRI	National Defense Research Institute
NDU	National Defense University
NPS	Naval Postgraduate School
O&S	operations and support
OCX	Operational Control System
OPM	U.S. Office of Personnel Management
OSD	Office of the Secretary of Defense
OUSD(A&S)	Office of the Under Secretary of Defense for Acquisition and Sustainment
PARCA	Office of Performance Assessments and Root Cause Analyses
PB	President's Budget
PDASD	principal deputy assistant secretary of defense
PE	program element
PEO	program executive officer

PII	personally identifiable information
PM	program management
PMRT	Project Management Resource Tools
POM	program objective memorandum
PQM	production, quality, and manufacturing
PROPIN	proprietary information
Q	question
R&D	research and development
RDAIS	Research, Development, and Acquisition Information System
RDT&E	research, development, test, and evaluation
ROI	return on investment
S&T	science and technology
SAE	service acquisition executive
SAR	Selected Acquisition Report
SBIR	Small-Business Innovation Research
SDW	Shared Data Warehouse
SEI	Software Engineering Institute
SET	Systems Engineering Transformation
SETA	systems engineering and technical assistance
SNaP-IT	Select and Native Programming—Information Technology
SQL	Structured Query Language
STTR	Small-Business Technology Transfer
T&E	test and evaluation
UARC	university-affiliated research center
URED	Unified Research and Engineering Database
USAF	U.S. Air Force
U.S.C.	United States Code
USD	under secretary of defense

USD(A&S)	Under Secretary of Defense for Acquisition and Sustainment
USD(AT&L)	Under Secretary of Defense for Acquisition, Technology, and Logistics
USD(R&E)	Under Secretary of Defense for Research and Engineering
VAMOSC	Visibility and Management of Operation and Support Costs

1. Introduction

Outline of the Report

What's in this Briefing

- Background
- Approach and methodology
- Findings on congressional questions
- Conclusions

The report is organized with four chapters:

- introductory background materials on data analytics
- approach and methodology
- findings on the questions from Congress
- conclusions.

Background

Congress has expressed strong interest in DoD's use of *Data Analytics* in DoD's decision-making over the last several years

- In particular, the Senate version of the FY 2017 NDAA contained a provision (Section 853) that would have mandated the establishment of activities to promote the use of data analytics and other evaluation-related methods to support acquisition decision-making, but the provision was dropped in the FY 2017 NDAA
- The conferees asserted:
  - *...a widespread recognition that the Department of Defense (DoD) does not sufficiently incorporate data into its acquisition-related learning and decision-making...These policies are sometimes based on assumptions, and program reviews do not always sufficiently incorporate relevant data against which to evaluate success...Government Accountability Office reported in 2015 that DoD officials responsible for acquisitions and developing requirements lacked access to data and the analytical tools necessary to conduct effective reviews...The conferees believe that data analysis and other evaluation-related methods are a critical element in making well-informed acquisition decisions and managing programs. As the Congressional Research Service noted, a lack of data or effective data analyses can lead to incorrect or misleading conclusions. The result may be policies that squander resources, waste taxpayer dollars, and undermine the effectiveness of government programs or military operations.*

    Joint Explanatory Statement of the Committee of Conference (H.R. 114-840, pp. 1125–1126) Accompanying the National Defense Authorization Act (NDAA) for fiscal year (FY) 2017, S. 2943
- Our briefing supports the Secretary's briefing as directed by the Conference Committee.
- A similar measure to the unenacted Senate section was passed in Section 913 of the FY 2018 NDAA that resulted in a directive to "establish a set of activities that use data analysis, measurement, and other evaluation-related methods to improve the acquisition outcomes of the Department of Defense and enhance organizational learning"

SOURCES: U.S. House of Representatives, 2016, pp. 1125–1126; Pub. L. 114-328, 2016; Pub. L. 115-91, 2017.
NOTE: FY = fiscal year; NDAA = National Defense Authorization Act.

Congress has expressed strong interest in the use of data analytics in the U.S. Department of Defense's (DoD's) decisionmaking during the past several years. In particular, the unenacted Senate version of the FY 2017 NDAA contained a provision (Section 853 of the June 14, 2016, version of S. 2943 that passed the Senate but was dropped by the House of Representatives) that would have mandated the establishment of activities to promote the use of data analytics and other evaluation-related methods to support acquisition decisionmaking, but the provision was dropped in the FY 2017 NDAA (Pub. L. 114-328, 2016) in lieu of direction in the Joint Explanatory Statement of the Committee of Conference for a briefing from the Secretary of Defense on data analytics. For context, the conferees asserted in the committee report that there was

> a widespread recognition that the Department of Defense (DoD) does not sufficiently incorporate data into its acquisition-related learning and decisionmaking. . . . These policies are sometimes based on assumptions, and program reviews do not always sufficiently incorporate relevant data against which to evaluate success. . . . Government Accountability Office reported in 2015 that DoD officials responsible for acquisitions and developing requirements lacked access to data and the analytical tools necessary to conduct effective reviews. . . . The conferees believe that data analysis and other evaluation-related methods are a critical element in making well-informed acquisition decisions and managing programs. As the Congressional Research Service noted, a lack of data or effective data analyses can lead to incorrect or misleading conclusions. The

result may be policies that squander resources, waste taxpayer dollars, and undermine the effectiveness of government programs or military operations. (U.S. House of Representatives, 2016, pp. 1125–1126)

A similar measure to Section 853 of the unenacted Senate version of the FY 2017 NDAA was passed as Section 913 of the FY 2018 NDAA, which resulted in a directive that "the Secretary of Defense shall establish a set of activities that use data analysis, measurement, and other evaluation-related methods to improve the acquisition outcomes of the Department of Defense and enhance organizational learning" (Pub. L. 115-91, 2017). This enacted statute essentially requires the implementation of the activities that the 2017 conference committee directed to be briefed.

Our briefing provides analysis in support of the secretary's briefing, as directed by the 2017 committee report.

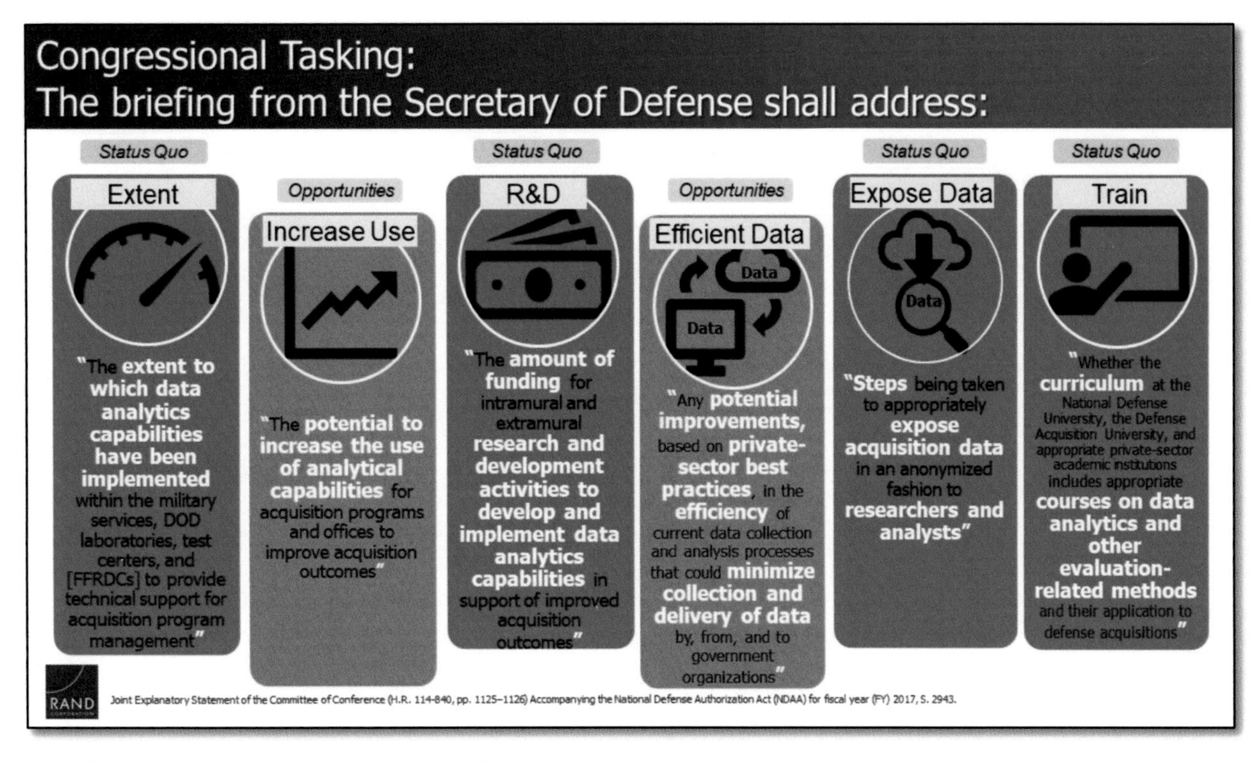

SOURCE: U.S. House of Representatives, 2016, pp. 1125–1126.
NOTE: FFRDC = federally funded research and development center; R&D = research and development.

The scope of Congress's inquiry is the use of any type of relevant analytics and other evaluation-related methods for acquisition that utilize data, the enabling of data management (e.g., best practices for data collection, delivery, and availability), and opportunities to improve these aspects as they pertain to improving acquisition program management and decisionmaking. Congress directed that the briefing address four topics about the status quo and two types of opportunities (U.S. House of Representatives, 2016, p. 1126). We framed these as the following research questions that align directly with Congress's six topics:

1. **Current extent of implementation:** What is the extent to which data analytics capabilities have been implemented across the DoD to provide technical support for acquisition program management?
2. **Potential to increase use:** What is the potential to increase this use of analytical capabilities?
3. **Current R&D funding:** What is the current amount of R&D funding for acquisition data analytics capabilities?
4. **Potential improvements in data collection and analysis:** What private-sector best practices could be leveraged to minimize the collection and delivery of data?
5. **Current data-sharing efforts:** What steps are being taken to share anonymized acquisition data to researchers and analysts?
6. **Current data analytics training:** Are the curricula at defense acquisition workforce (AWF) training institutions including appropriate courses on applied and general data analytics and other evaluation-related methods?

This report summarizes our analysis on these six topics (which we framed as questions), including perspectives to help Congress understand the larger context and realities of the findings, as well as opportunities and suggestions for Congress and the DoD.

Congressional changes have resulted in an uncertain environment for acquisition information governance, management, and use

- Recent NDAA changes in defense acquisition roles, responsibilities, and organizational structure within DoD are affecting data governance, management, and use
  - Section 825 of the FY 16 NDAA delegated decision-making to the Service Acquisition Executives (SAEs) for new major defense acquisition programs (MDAPs)
  - Section 901 of the FY 17 NDAA eliminated the USD(AT&L) and created an USD for Research and Engineering [USD(R&E)] and the new USD(A&S) along with a Chief Management Officer (CMO)
- Additional legislation has further complicated data governance and management
  - The submission of a SAR to Congress for major acquisition programs was repealed by the FY 2018 NDAA, effective December 31, 2021 (See Public Law (P.L.) 115-91, Section 1051(x)(4), signed into law on December 12, 2017). Congress is expecting a proposal for an improved SAR
- These changes have led to concerns about
  - What data is needed for the new roles in OSD and the Services
  - The loss of longitudinal program data to analyze for policy reasons as well as strategic questions about analysis to conduct

SOURCES: Pub. L. 114-92, 2015; Pub. L. 114-328, 2016; Pub. L. 115-91, 2017.
NOTE: SAE = service acquisition executive; MDAP = Major Defense Acquisition Program; USD(AT&L) = Under Secretary of Defense for Acquisition, Technology, and Logistics; USD = under secretary of defense; USD(R&E) = Under Secretary of Defense for Research and Engineering; USD(A&S) = Under Secretary of Defense for Acquisition and Sustainment; CMO = chief management officer; SAR = Selected Acquisition Report; OSD = Office of the Secretary of Defense.

Recent NDAAs have significantly changed defense acquisition roles, responsibilities, and organizational structure within the DoD. Section 825 of the FY 2016 NDAA delegated decisionmaking to the SAEs for new MDAPs (see 10 U.S.C. 2440, as amended; Pub. L. 114-92, 2015), while Section 901 of the FY 2017 NDAA eliminated the USD(AT&L) position and created the USD(R&E) and the new USD(A&S), along with a CMO (Pub. L. 114-328, 2016). In addition to these major leadership changes and their associated supporting organizational changes, some legislation directly affects data management. For instance, the submission of a SAR to Congress for major acquisition programs was repealed by the FY 2018 NDAA, effective December 31, 2021 (see Pub. L. 115-91, 2017, Section 1051(x)(4), signed into law on December 12, 2017). The current absence of data governance over such information creates uncertainties regarding the authoritative sources, definitions, and standards of specific data, impeding collection and limiting access and use. The DoD is still working out the details of information governance and management in this emerging acquisition environment, but given the new roles of the acquisition organizations within the DoD, failure to determine what data are needed is risky.

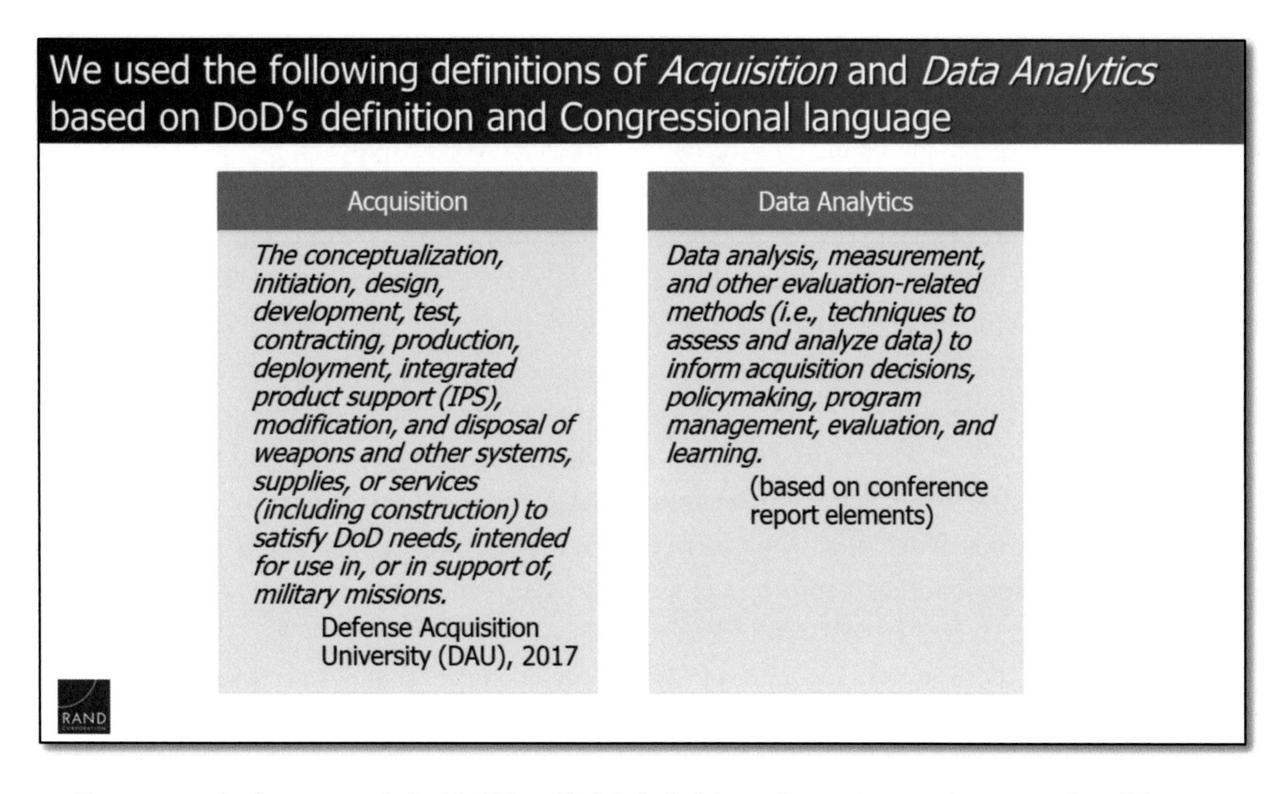

For our analysis, we used the DoD's official definition of *acquisition* shown on the slide as published by DAU (2017a). For what constitutes *data analytics* for our study, we constructed the definition shown on the slide based on our general reading of the conference report (U.S. House of Representatives, 2016, pp. 1125–1126) and our own knowledge of the field.

We included—but did not limit—our review to "advanced" analytics

- "Advanced" is a continuum that varies by domain
  - e.g., simpler statistics and advanced statistical techniques
- What matters is using the right techniques to inform decisions, not the label
- More advanced techniques:
  - May not be applicable to particular data or questions
  - Require specialized expertise
- We assessed DoD's movements to advances as well as useful simpler approaches that improve decisions

Examples of Analytic Sophistication

(Increasingly Advanced ↑)

Business Intelligence	Statistics	Systems Engineering	Supply-Chain Analysis
Predictive modeling and analysis	Principal-Component Analysis, Deep Neural Networks	Model-based design, requirements, and multi-system analysis	Network dependency and graph analysis with foreign dependencies
Integrated data, aggregation, and visualization tools	Multi-variate regression	System-of-system analysis	Multi-tier supplier reviews
Preset calculations and dashboards	Single-variate regression	CAD design reviews	Assessing first-tier suppliers
Custom simple calculations	Simple averages and trends	Paper-based interface and design analysis	Identifying sole suppliers

SOURCE: RAND assessment

NOTE: CAD = computer-aided design.

Given all the press and public breakthroughs in more-advanced analytics, it is important to note that we included—but did not limit—our review to advanced analytics for a number of reasons. *Advanced* is really a continuum, not a clearly separable category of techniques (see, for example, the table in the slide above illustrating the range of increasingly advanced techniques in the fields of business intelligence, statistics, systems engineering, and supply-chain analysis). For example, even in the field of statistics, the complexity of techniques ranges from simpler to more advanced. What matters is using the right techniques to inform decisions, not an arbitrary label or techniques (e.g., "big data" or artificial intelligence, including machine learning and neural networks) that are currently receiving much publicity. Also, more-advanced techniques might not be applicable to a particular set of data or a particular policy question or decision. In addition, more-advanced techniques often require specialized expertise, equipment, data, and investments, so applications are limited to critical questions.

Thus, we assessed the DoD's movements toward more-advanced (but not necessarily the most-advanced) capabilities based on the applicability of the techniques. We also reviewed the full range of simpler approaches that, when applied to acquisition, can improve decisions. This approach also aligns with Congress's questions, since they explicitly asked not only about data analytics but also about other evaluation-related methods.

Review of Industry Best Practices

Why data governance, management, and analytics are important competencies for an organization...

McKinsey Analytics:
"Data is now like air. It's all around us. It has become common knowledge that the world churns out an enormous and expanding amount of data each day—billions of gigabytes... In industry, all organizations create data, and as storage costs continue to tumble, more of it is being kept and analyzed to create competitive advantages. Increasingly, organizations also share data with other companies to realize new businesses, giving rise to the beginnings of digital ecosystems that stand poised to blur traditional industry borders" (McKinsey Analytics, 2018, p. 3).

Harvard Business Review:
"Cross-industry studies show that on average, less than half of an organization's structured data is actively used in making decisions—and less than 1% of its unstructured data is analyzed or used at all. More than 70% of employees have access to data they should not, and 80% of analysts' time is spent simply discovering and preparing data. Data breaches are common, rogue data sets propagate in silos, and companies' data technology often isn't up to the demands put on it" (DalleMule and Davenport, 2017).

MIT:
"Organizations across the business spectrum are awakening to the transformative power of data and analytics. They are also coming to grips with the daunting difficulty of the task that lies before them. It's tough enough for many organizations to catalog and categorize the data at their disposal and devise the rules and processes for using it. It's even tougher to translate that data into tangible value. But it's not impossible, and many organizations, in both the private and public sectors, are learning how" (MIT Sloan Management Review, 2016).

IBM Research:
"Much has been written about the explosion of data, also known as the "data deluge". Similarly, much of today's research and decision making are based on the de facto acceptance that knowledge and insight can be gained from analyzing and contextualizing the vast (and growing) amount of "open" or "raw" data. The concept that the large number of data sources available today facilitates analyses on combinations of heterogeneous information that would not be achievable via "siloed" data maintained in warehouses is very powerful." (Terrizzano, et al., 2015, p. 1).

Both IBM and McKinsey describe the vast amount of information being collected today. IBM calls it an "explosion of data, also known as the 'data deluge'" (Terrizzano et al., 2015, p. 1). McKinsey Analytics notes that "the world churns out an enormous and expanding amount of data each day. . . . In industry, all organizations create data, and as storage costs continue to tumble, more of it is being kept and analyzed to create competitive advantages" (McKinsey Analytics, 2018, p. 3). As is the case in the private sector, the DoD is generating an enormous amount of data. These data are collected and used for many purposes. This report focuses on the information created and used during DoD acquisition and how those data can be utilized through analysis to improve decisionmaking.

The *Harvard Business Review* has also identified several challenges in industry: "[L]ess than half of an organization's structured data is actively used in making decisions—and less than 1% of its unstructured data is analyzed or used at all [and] 80% of analysts' time is spent simply discovering and preparing data. Data breaches are common, rogue data sets propagate in silos" (DalleMule and Davenport, 2017). Other recent RAND Corporation analyses observed similar challenges in DoD acquisition data access and management (see Riposo et al., 2015; McKernan et al., 2016, 2017, 2018).

Finally, the Massachusetts Institute of Technology (MIT) describes additional challenges involved in governing and managing information: "It's tough enough for many organizations to catalog and categorize the data at their disposal and devise the rules and processes for using it. It's even tougher to translate that data into tangible value" (*MIT Sloan Management Review*,

2016). Given the magnitude of information being collected and stored by various organizations within the DoD, it has also been a challenge to translate these data into tangible value. Although identifying the questions that DoD decisionmakers need to answer should be a priority to help understand the value of the DoD's data, that step is sometimes lost within the DoD's data collection efforts.

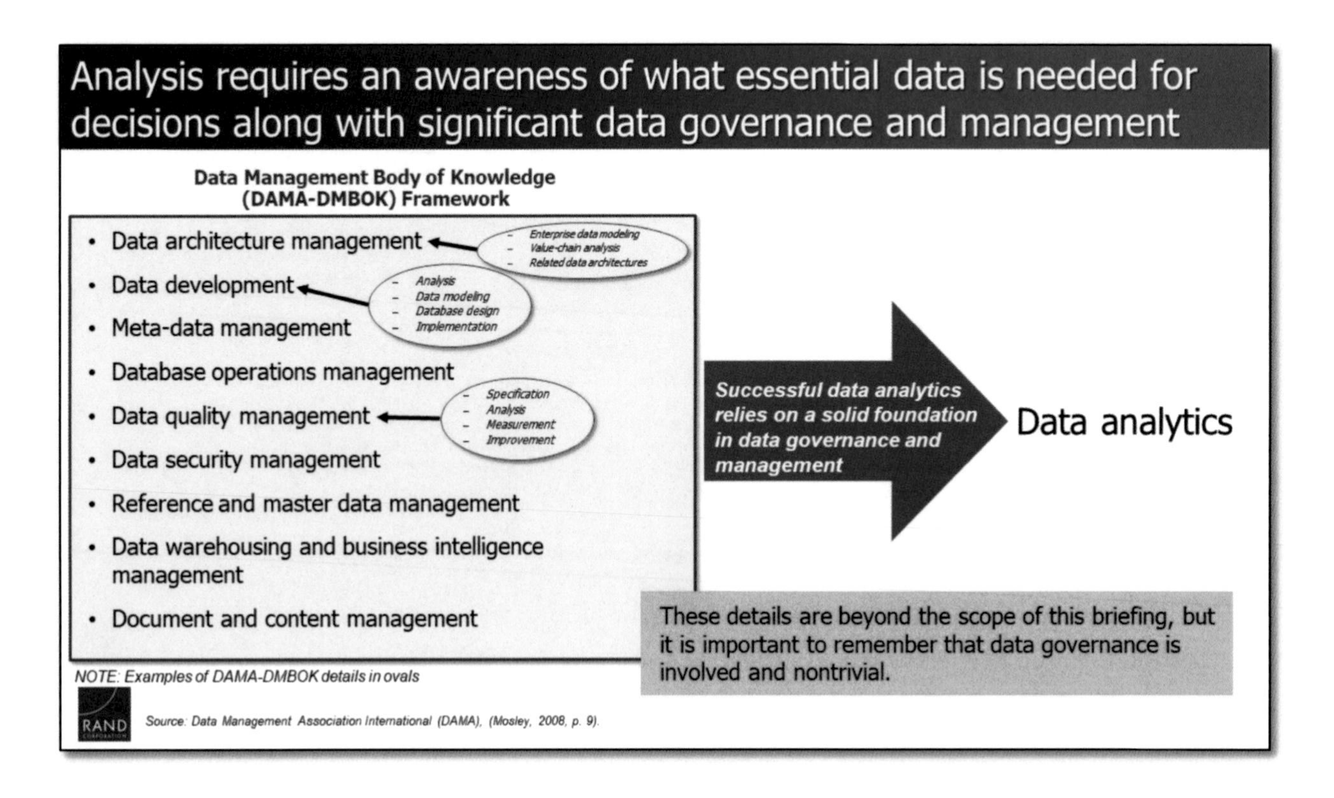

NOTE: DAMA-DMBOK = Data Management Association International's Data Management Body of Knowledge.

For both the public and private sector to conduct widespread analytics, organizations need to implement sound data governance and management practices upfront. The complexity of data governance and management cannot be understated. A significant amount of effort needs to be expended in such areas as data security, data-architecture management, data-quality management, and workforce training to implement successful analytics.

Many organizations in the private sector have built models and instructions for successful data governance and management practices. For instance, the DAMA-DMBOK framework is shown above. It defines *data management* as the development, execution, and supervision of plans, policies, programs, and practices that control, protect, deliver, and enhance the value of data and information assets. A coherent data analytics plan means that resources have to be properly identified. As more data need to be processed, more resources are necessary. Using a data governance framework, such as the DAMA-DMBOK, is a useful and accepted method of standardizing data analytics planning, to better identify what needs to be done, and what resource requirements will be (Mosley, 2008, p. 9). The framework includes architecture management, development, operations management, security management, warehousing, and quality management. Selected subtasks (including some analysis) are shown in ovals in the slide.

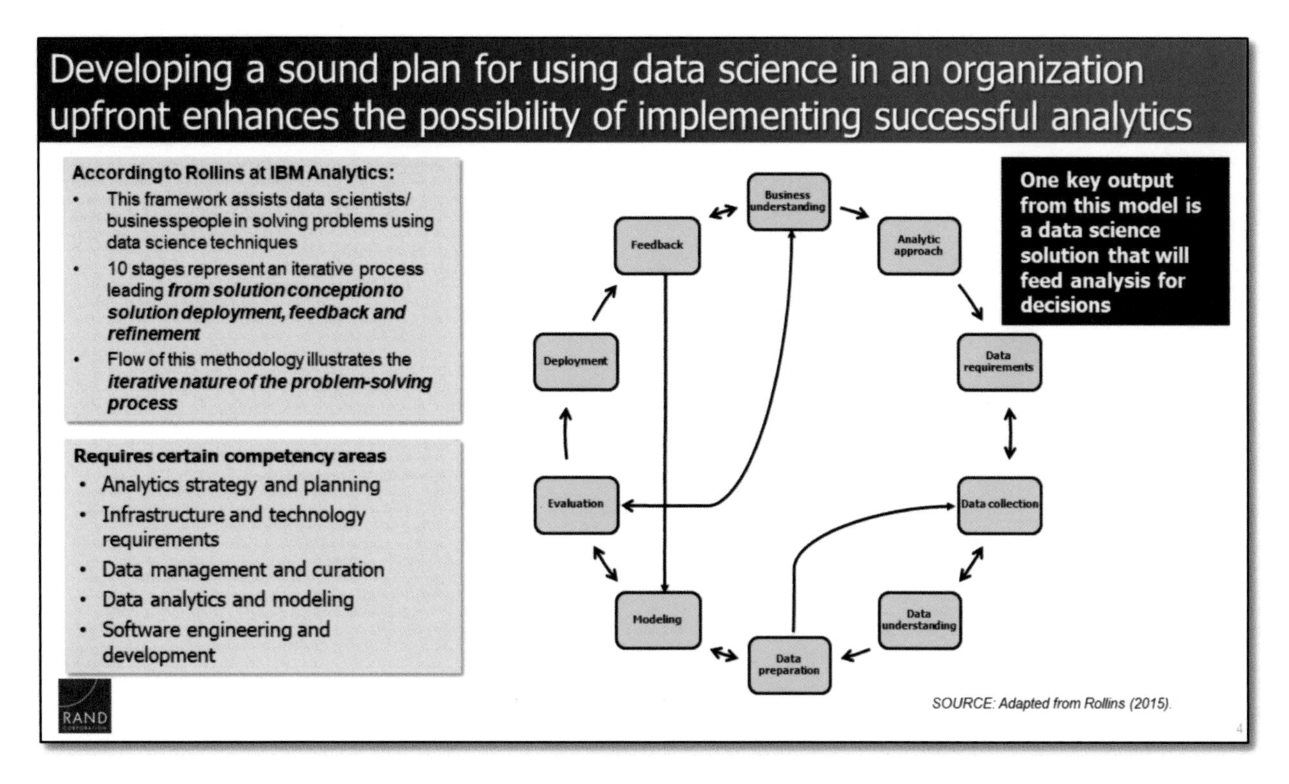

Others in the private sector have also identified the need for frameworks or models that guide organizations to successfully integrate data analytics. An additional framework developed by Rollins at IBM assists data scientists or businesspeople who want to solve problems using data analytics techniques (Rollins, 2015). This framework provides an iterative process to help with planning for successful data analytics implementation in organizations. This process can be conceptualized as a loop. To begin the initial process, an understanding of the business (or, in the case of the DoD, understanding the purpose of the analysis) is necessary. This is where analysts work with stakeholders to gain an understanding as to why a data analytics solution is needed. Using the knowledge gained from the first step, data scientists or functional experts develop an analytic approach. Identification, collection, understanding, and preparation of data are the next four steps, which can often feed back to each other as the available data become clearer. These four steps can often take the majority of the time during the analysis. Once the data are prepared, modeling, evaluation of the model outputs, and deployment of the lessons gathered from the data analysis are implemented. A final step, feedback, is when the stakeholders can give further insights, which allows the analysts to begin the process again to further tune the analysis. User feedback may result in changes to the database that must be planned, analyzed, designed, developed, tested, implemented, and, then, maintained (Eternal Sunshine of the IS Mind, 2013). This is also the step when analysts perform maintenance on the data and associated data systems. Maintenance of a piece of software, data, or analysis plan is often overlooked, because it is considered to be a less glamorous step (Kay, 2002). Once a data analytics process is developed and operational, the job of maintaining it is key to continued operational viability, relevance, and data quality.

2. Approach and Methodology

What's in this Briefing

- Background
- Approach and methodology
- Findings on congressional questions
- Conclusions

Research approach addresses the breadth of Congressional topics: data, analysis, and improvement opportunities

- Inputs:
  - Discussions with select stakeholders
  - Existing literature, public budget exhibits, policies, legislation, training course catalogs and curricula; inventories of relative IT systems
  - Experience, knowledge, and judgement of authors, as noted
- Study techniques
  - Functional decomposition and mapping; workforce cost analysis; budget and solicitation analysis; literature review and analysis; analysis of industry best practices; analysis of DoD data management and analytic activities.
- Assumptions
  - Data analytics are a means for improving decisions
    - *We lay out analytics that support specific decisions, but we generally do not directly study cause-and-effect*
  - Advanced business analytics have some DoD applications
    - *Found some application and explorations, but further applications require continued investigation*

Approach

Our study employed a variety of approaches to answer Congress's six questions. In particular, we used the decomposition and review of acquisition into functions as a means of understanding how data and associated analytics and other evaluation-related methods are used to support acquisition decisions. The breadth of sources and their seemingly diverse nature reflect the breadth of the questions and the fact that activities, capabilities, and budgets are often not tracked by data analytics and enabling data feeds.

The breadth of questions and limits of data drove us to use a wide range of analyses to gain some perspective on the answers. These are outlined briefly on the slide. We also assumed (rather than studied in depth) that there may be uses for recent advanced analytics (particularly those used by the commercial business sector) in the DoD; we found some applications, but, given the breadth of Congress's topics, we did not focus solely on such data analytics.

Scope and Limitations

In assessing the extent of data analytics capabilities, we explore not only what data, analytic capabilities, and analytic results exist in defense acquisition but also how they are used.

It is important to note that the definitions of *acquisition* and *data analytics* scope our study. The broad definition of *acquisition* means that we did not limit our study to development and procurement but included subsequent activity, such as sustainment. It also is important to note that, in focusing on acquisition, we did not address analytics for related areas that affect acquisition, including force planning, requirements, budgeting and resourcing, intelligence, and military operations.

Our definition of *data analytics* implies that we did not limit our study to what might be termed *advanced* analytics but instead examined what Congress asked about: the full range of data analytics and other evaluation-related methods. Advanced analytics for acquisition could be deemed, for example, the application of big data, machine learning, and predictive analytics to various acquisition data. Advanced analytics could also be more-mature developments, such as transforming systems engineering to complex model-based data analysis and design systems. We did, though, seek to assess what advanced techniques are being pursued and what kind of applications they may be appropriate for. We also present examples of data analytics as a continuum to illustrate any advancement made by the DoD and how simpler analytics can still be useful, but we do not spend time trying to define a strict set of levels in that continuum.

Although we examine the breadth of data analytics and enabling data for major acquisition decisions (e.g., Milestone B) and in information systems that archive and make available various data, we did not assess in detail what specific data analytics and enabling data are needed and not needed. For example, we did not assess the specific minimum data needed at various levels in the acquisition chain of command for analysis and oversight.

For inputs to our study, a data call was initially proposed as a mechanism to collect information about Congress's six questions, but it was deemed infeasible and would likely not have produced consistent data that could be assessed DoD-wide. Also, instead of asking the DoD components to assess available information about data analytics, we and our sponsors decided to examine those data ourselves to both reduce burden on the DoD components and pursue a more consistent methodology. Author experience, knowledge, and judgment were employed in our study to assess and synthesize available information and fill gaps in primary and secondary inputs. Research for this study was conducted by the following team members.

- Philip S. Anton is a senior information scientist who conducts research on acquisition and sustainment policy, cybersecurity, emerging technologies, technology foresight, process performance measurement and efficiency, data science and analytics, aeronautics test infrastructure, and military modeling and simulation.
- Megan McKernan is a senior defense researcher with more than 14 years of experience conducting DoD acquisition analyses. She is co-leading research examining DoD acquisition data management.
- Ken Munson is a senior management scientist specializing in acquisition, logistics, budget, requirements, and policy analysis.
- James G. Kallimani is a senior technical analyst with an engineering and data analytics background, specializing in problem-solving and creative thinking for public policy, national security, defense, and government.
- Alexis Levedahl is a research assistant with experience in emerging technologies, acquisition, and requirements development.

- Irv Blickstein is a senior engineer with 50 years of experience in the field of defense analysis and management with a specialty in planning, programming, and budgeting, as well as acquisition.
- Jeffrey A. Drezner is a senior policy researcher who has conducted policy analysis on a wide range of issues, including energy R&D planning and program management, best practices in environmental management, analyses of cost and schedule outcomes in complex system development programs, aerospace industrial policy, defense acquisition policy and reform, and local emergency response.

Finally, there are other research questions regarding data analytics and enabling data that we could have studied but were out of the project's scope—e.g., what data are most valuable, and how can we understand costs and benefits; how can we assess the appropriateness of data usage; what specific approaches can be used to counter the culture of not sharing data; what is the optimal mix of advanced data analytics commercial off-the-shelf (COTS) tools, and simpler analytic approaches; is the extent of data analytics capabilities sufficient or appropriate (how much spending is appropriate for data analytics—against what baseline and benchmarks); and what is the quality of the data analytics courses at defense training institutions? Some of these research topics are raised later as possible next steps.

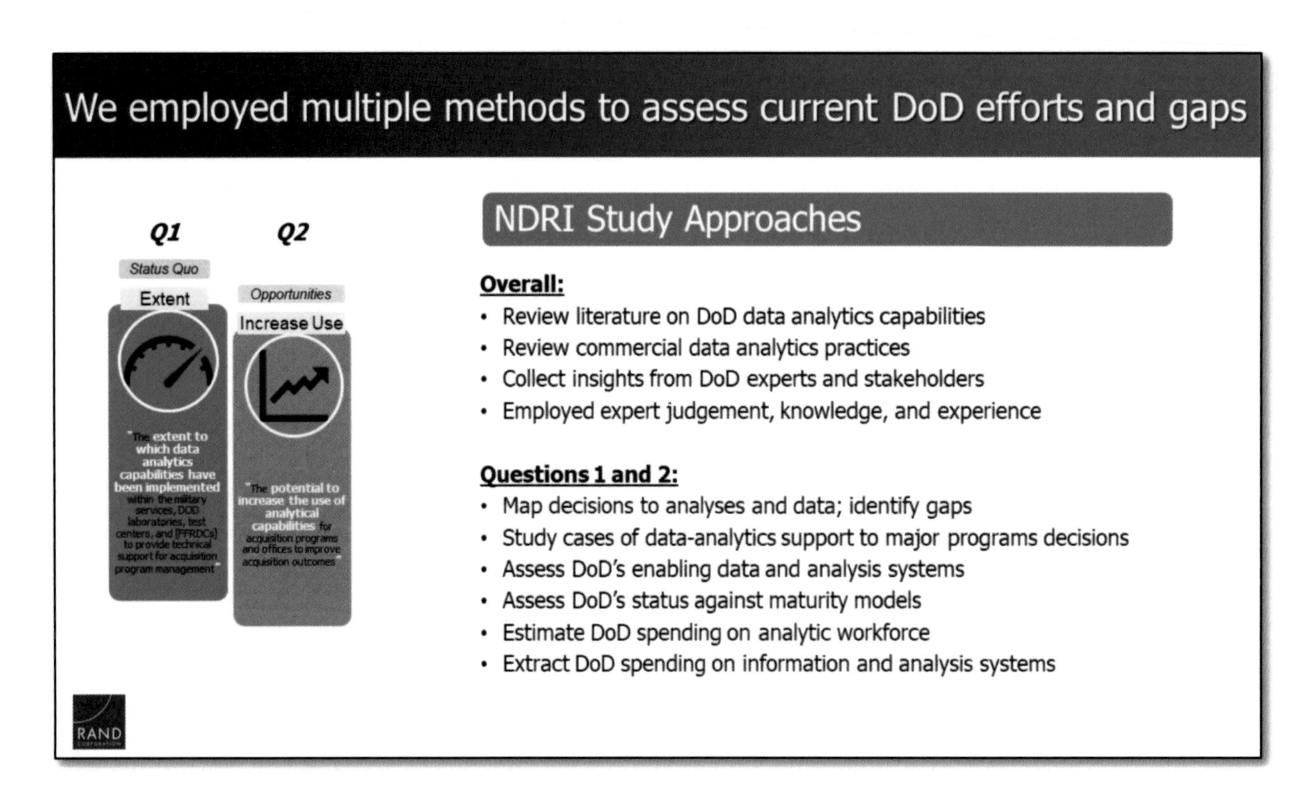

NOTE: NDRI = National Defense Research Institute.

Given the breadth and scope of Congress's questions, we used different approaches to assess the current status and opportunities for improvement. On the above slide and the next one, we map each question to the basic approach taken. Overall, we reviewed the literature on the DoD's data analytics capabilities, including internal documentation, and we collected additional information from DoD experts and stakeholders. We also reviewed commercial data analytics best practices, methodologies, and advice.

For Congress's first two questions, about the extent of data analytics and the potential to improve data analysis, we used six approaches to obtain as broad a view as possible and avoid simply reporting examples of DoD data analytics efforts:

1. Because Congress's concern is improving acquisition program management and decisionmaking, we mapped decisions (primarily from milestone B) to analyses that supported those decisions and the data used for those analyses. This mapping allowed us both to assess the breadth of data analytics across acquisition functions and to identify gaps and challenges.
2. We examined existing analyses of 13 programs with documented program decision challenges to assess the extent and availability of data analytics and the decisionmakers' use of those data analytics.
3. We assessed the status and evolution of the DoD's data and analysis systems.
4. We examined established and evolving data analytics maturity models from industry and the federal government and assessed the DoD's status against these models.
5. We estimated the DoD's spending on the portions of the AWF that are primarily analytic, as a way to quantify the extent of personnel-based data analytics capabilities.
6. We summarized the annual spending on major IT systems (not desktop computing) supporting acquisition, as reported in detailed budget exhibits.

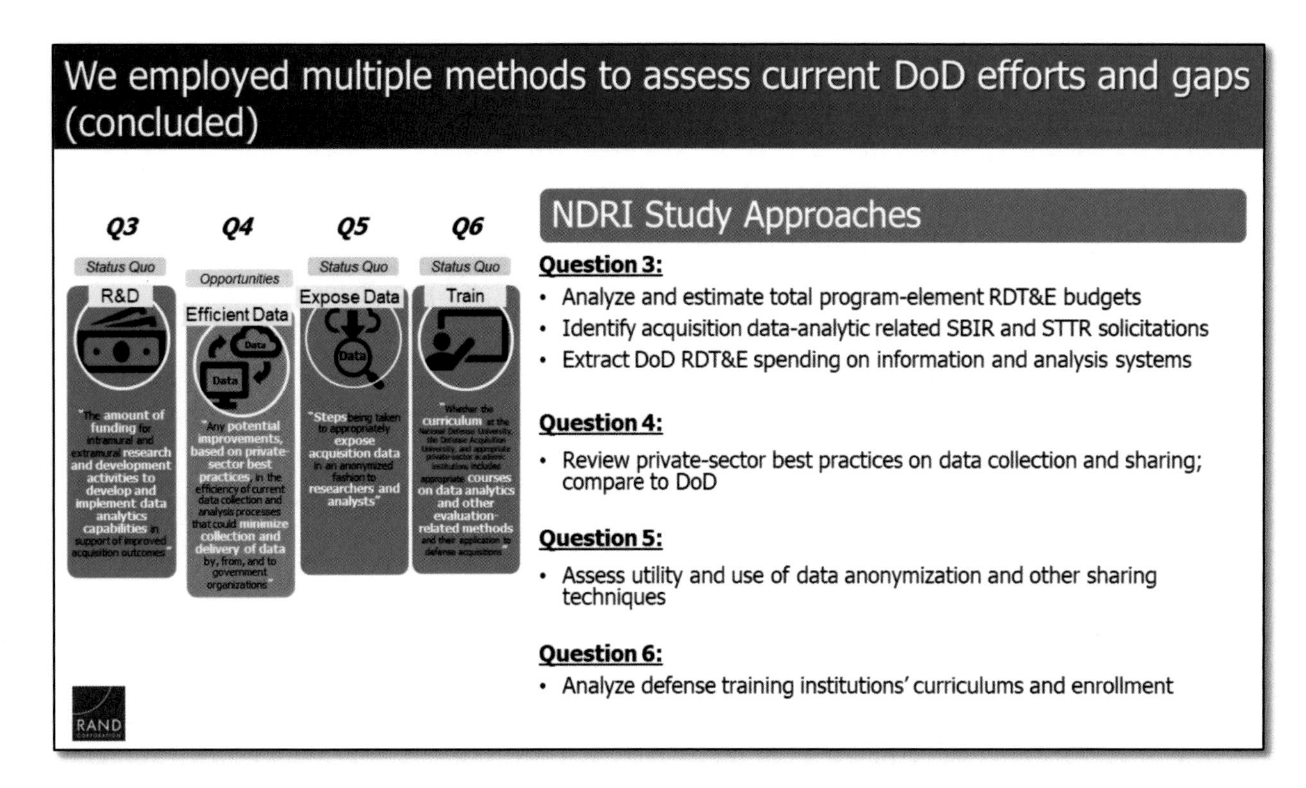

NOTE: RDT&E = research, development, test, and evaluation; SBIR = Small-Business Innovation Research; STTR = Small-Business Technology Transfer.

For Congress's third question, about the amount of R&D funding to develop acquisition data analytics capabilities, we reviewed the published descriptions of each RDT&E program element (PE), both DoD-wide and for each military department in the President's Budget (PB) request for FY 2019. We also reviewed the recent SBIR and STTR solicitations to identify any data analytics topics related to acquisition. In addition, we extracted the RDT&E portions of the annual spending on major IT systems (not desktop computing) that support acquisition, as reported in detailed budget exhibits.

To address Congress's fourth question, we reviewed the private sector's best practices for efficient data collection and delivery, then compared those practices with those either in place now or being pursued by the DoD.

For Congress's fifth question, we directly reviewed DoD steps being taken to appropriately expose acquisition data in an anonymized fashion to researchers and analysts. We also assessed the utility of anonymization as a sharing mechanism. Alternatives include approaches assessed for question 4.

Finally, in response to Congress's training question (question 6), we reviewed the curricula at the DoD's primary defense acquisition training institutions and several partner institutions. This review involved reading the descriptions of many hundreds of courses in these institutions' catalogs and assessing whether data analytics are addressed and applied. We also discussed content and tools with the Defense Acquisition University (DAU)—the DoD's primary training house for the AWF. Finally, we obtained enrollment data for DAU courses, to help understand penetration across the AWF. Given the scope and tasking, we did not assess the quality of the data analytics training.

3. Findings on Congressional Questions

Questions 1 and 2: Extent of, and Potential to Increase, DoD Data Analytics Capabilities

Congress's first two topic areas asked about

- "the extent to which data analytics capabilities have been implemented within the military services, DOD laboratories, test centers, and [FFRDCs] to provide technical support for acquisition program management" (U.S. House of Representatives, 2016, p. 1125)
- "the potential to increase the use of analytical capabilities for acquisition programs and offices to improve acquisition outcomes" (U.S. House of Representatives, 2016, p. 1125).

We addressed these topics together because our analyses related to *extent* also identified gaps and opportunities related to the potential to increase the DoD's data analytics capabilities.

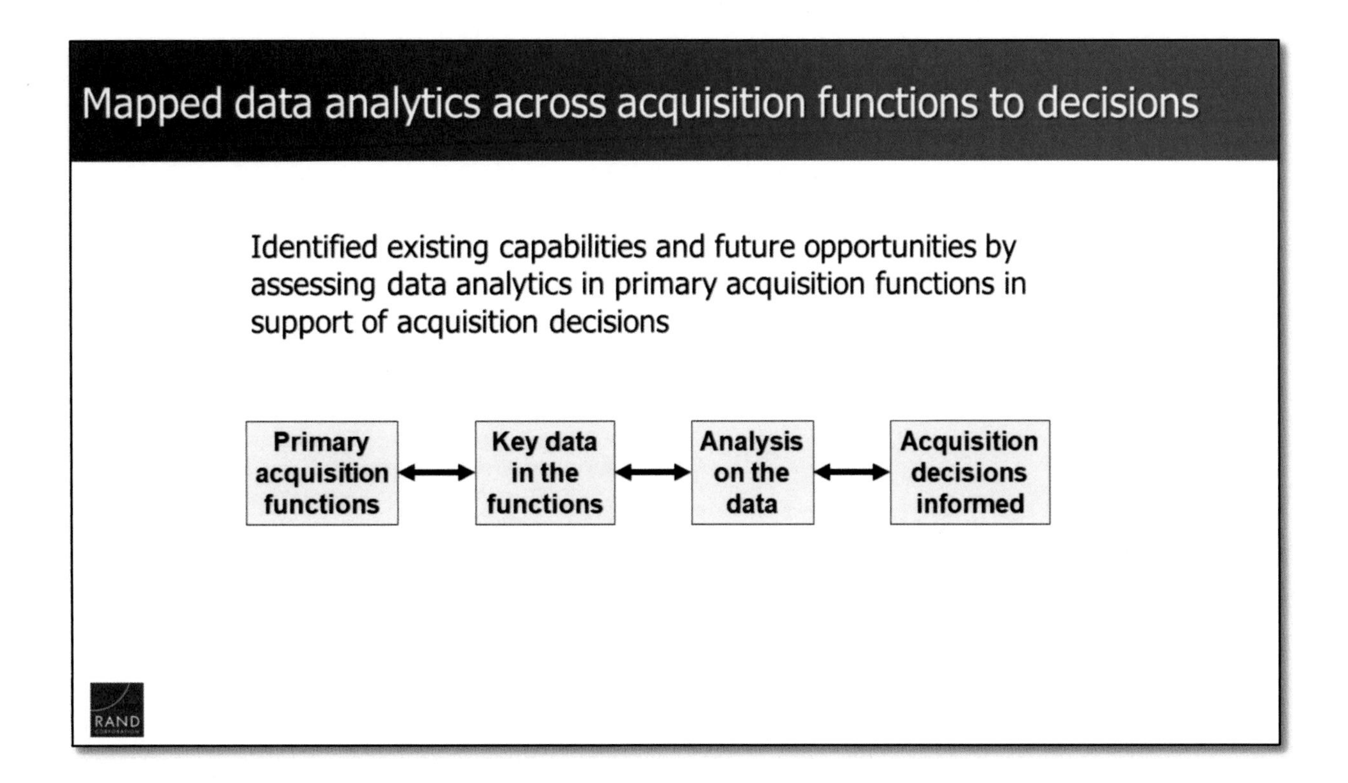

Map Decisions to Analyses and Data, Then Identify Gaps

One approach we used to understand the extent of data analytics in support of acquisition decisions and program management is to review the full range of data analyses across all acquisition functions as they relate to specific types of acquisition decisions. Here, the use and application of data analytics we assessed are those executed through major acquisition functions. The slide above illustrates the conceptual map from functions to data within the functions, to analysis about those data, and ultimately to the acquisition decisions that are informed. For this map, we leveraged our expertise and reviewed relevant statutes, DoD policy and guidance on acquisition (key functions, decision points, information and data elements, and major supporting analysis), proposed or unimplemented approaches, and commercial analogs.

These primary government functions enable the acquisition of goods and services

Primary Acquisition Functions

1. Program Management/Manager
   - 1.1 Business case and economic analysis
   - 1.2 Affordability analysis
   - 1.3 Acquisition strategy
   - 1.4 Risk management
   - 1.5 Technical maturity
   - 1.6 Personnel and team management
   - 1.7 Business and marketing practices
   - 1.8 configuration management
2. Research and Development (R&D)
3. Engineering
   - 3.1 Systems engineering
   - 3.2 Facilities engineering
   - 3.3 Software/IT
4. Intelligence & Security
   - 4.1 Cybersecurity
   - 4.2 Program Protection
5. Test and Evaluation (T&E)
   - 5.1 Developmental T&E
   - 5.2 Operational T&E
6. Production, Quality, and Manufacturing (PQM)
7. System and Operational Issues
   - 7.1 Spectrum (frequency allocation, emissions, etc.)
   - 7.2 Environmental
   - 7.3 Energy
8. Product Support, Logistics, and Sustainment
9. Financial Management
10. Cost Estimating
11. Auditing
12. Contract Administration
    - 12.1 Contracting actions
    - 12.2 Contracting strategy
    - 12.3 Contract peer review
    - 12.4 Acceptance of deliverables
13. Purchasing
14. Industrial Base and Supply-Chain Management
15. Infrastructure and Property Management
16. Manpower Planning and Human Systems Integration
17. Training and education
    - 17.1 Training and education for government execution
    - 17.2 Training and education for acquired systems
18. Disposal

Acquisition Interface Functions

19. Requirements: receive, inform, and fulfill
20. Acquisition Intelligence: request, receive, and respond
21. Legal Counsel: request and act upon

RAND

SOURCES: Anton et al (forthcoming), DAU (2002, 2004), Parker (2011), MCSC (2017), NAVAIR (2015), OPM (2009), USD(AT&L) (2016b)

NOTE: T&E = test and evaluation; PQM = production, quality, and manufacturing; MCSC = Marine Corps Systems Command; NAVAIR = Naval Air Systems Command; OPM = U.S. Office of Personnel Management.

First, we used the functional decomposition of DoD acquisition and procurement (broadly defined) shown in the slide above. This taxonomy of acquisition functions is based on Anton et al.'s (forthcoming) review of acquisition instructions, guidebooks (DAU, 2002, 2004; Parker, 2011; MCSC, 2017; NAVAIR, 2015), acquisition career fields (OPM, 2009; USD[AT&L], 2016b), and our experience in acquisition. As with any portfolio or taxonomy exercise, other lists can be developed, but this approach aligns well with the common functions as reflected generally in defense organizations, career fields, and acquisition documents.

Wide breadth of DoD analytics is generally available to support acquisition functions and decisions

Data Management (examples)

- Strategy, planning, and budgeting
- Data requirements and use cases
- Authoritative sourcing
- Collection, archiving, curation, and sharing
- Security
- Backup and recovery
- Training and support
- Data definitions and standards
- Assessment, auditing, cleaning, and transformation
- Purging

Key Enabler

Analytic Techniques (Examples)

General:

- Quantitative analysis, qualitative analysis, data mining, statistical analysis, machine learning (e.g., classification, clustering, outlier detection, filtering), semantic analysis (e.g., natural language processing, text analytics , sentiment analysis, and visual analysis (e.g., heat maps , time series plots, network graphs, spatial data mapping), etc.

Applied:

- Market research, requirements analysis, mission analysis, technical maturity analysis, risk analysis and framing assumptions, DoD's risk-management framework, affordability analysis, cost estimating, schedule analysis, analysis of alternatives, tradeoff analysis, budget analysis, legal and policy analysis, engineering analysis, T&E, security, quality analysis, supply chain analysis, reliability and life-cycle sustainment analysis, contracting analysis, production analysis, auditing, sustainment analysis, etc.

Acquisition Decision-Making

- Congress
- Defense acquisition executive and others at OSD
- Component Acquisition Executive
- Program Executive Officer
- Program Manager

RAND

The slide above provides more detail on the breadth of DoD data analytics that support acquisition functions and decisions across the acquisition chain of command (the program manager, program executive officer [PEO], and SAE or the defense acquisition executive [DAE]) and other stakeholders (e.g., within OSD and in Congress).

A key enabler of analysis is data and associated management, shown on the left side of the slide. Besides the obvious point that analysis cannot be conducted without data, this exposé on data management and governance helps to illustrate that significant activity and investments are required to obtain and provide data of sufficient quality for analysis. Even advanced commercial data analytics efforts and best practices often involve significant data collection, cleaning, and preprocessing. One challenge is ensuring that data definitions are common to enable cross-organizational data analysis. In some cases, existing accounting standards and charts of account already provide such standardization for basic business-intelligence analysis, but, in other domains, commonality requires governance and active management. Sometimes data do not have to be standardized across all entities if the local definitions are available and can be transformed into a common standard, but that process requires at least some minimal level of data definitions. Finally, some advanced analytic tools can ingest unstructured data, but those applications may be limited, and such lack of structure imposes an added difficulty to the analysis. Generally, we found many cases of the DoD implementing these commercial best practices in data management and governance, but the level of maturity in these practices varied considerably across the DoD acquisition community.

The data then enable analyses, which can be separated into two categories. *General techniques* encompass a range of approaches that are independent of the specific application area. *Applied techniques* are specific (in our case) to various acquisition functions and analysis. This and subsequent slides illustrate what it means to *use* data analytics to inform acquisition decisions. *Use* is often implemented through functions that are related to various aspects of acquisition, not just in terms of the types of generic tools employed (e.g., Excel, Tableau, Qlik Sense, R, Stata, big data, and text understanding). In other words, an application is necessary when employing a tool, if that tool will inform a decision, and these slides help illustrate the breadth of that application for acquisition.

Acquisition decisionmaking requires a wide range of supporting analysis (Requirements, Technical, Risk, Budgets) *(Milestone B examples)*

Domain	Decision	Supporting Analyses	Analytic Capabilities				Gaps and Challenges
			DoD AWF	Other Resources	Analytic Approaches	Key Data: Available	
Require-ments	Design sufficient and meets requirements	Engineering analysis; Requirements and mission analysis	Engr.; PM	DoD labs; UARCs; FFRDCs; Requirements	Engineering and requirements analysis; wargaming	System design; requirements; threat envelops	Programs have requirements documents, but centralized info. system KM/DS archive is not always complete. Threat data may not be current.
	Requirements validated	Requirements and mission analysis; JCIDS Analysis: Functional Area Analysis (FAA), Functional Needs Analysis (FNA) and Functional Solutions Analysis (FSA)	PM; Leadership	Requirements	*(ref: JCIDS outside DAS)*		
Technical	Technology mature	Technical maturity and risk analysis	Engr.; IT; T&E	DoD labs, test centers, and FFRDCs	Risk analysis; Wide range of T&E analytic approaches and infrastructure		
	Market research; technology non-duplicative	Market research	Engr.; PM	DoD labs; UARCs, and FFRDCs	Internet search; technology and requirements analysis	Market and technical literature	Analyst access to PROPIN.
Risk	Risks acceptable given need	Framing assumptions; RMF	Engr.; IT; T&E; PM	DoD labs; UARCs, Test Centers; and FFRDCs	Framing assumptions; risk analysis; RMF	System concepts; risks	FA concept is new and little understood.
Budgets	Program affordable	Affordability analysis; cost analysis (costs within goals)	CE; PM	CAPE; Resource planners (G8, N8, A8); Leadership; FFRDCs	Budgeting systems, spreadsheets, optimization software	Budget elements; Component planning total (TOA); Cost data	Difficulty exists in mapping budget PEs to specific acquisition programs and detailed acquisition tasks

RAND

SOURCES: DoDI 5000.02; 10 U.S.C.; DAU (2002, 2004, 2017b), Parker (2011), MCSC (2017); NAVAIR (2015); plus author expertise and experience on analytic capabilities, gaps, and challenges.

NOTE: JCIDS = Joint Capabilities Integration and Development System; RMF = Risk Management Framework; engr. = engineer; PM = program management; CE = cost estimating; UARC = university-affiliated research center; CAPE = Cost Assessment and Program Evaluation; DAS = Defense Acquisition System; TOA = Total Obligational Authority; KM/DS = Knowledge Management/Decision Support; PROPIN = proprietary information; FA = framing assumption; DoDI = Department of Defense Instruction.

In the second step of our analysis, we reviewed key decisions at Milestone B (arguably the most important decision point in a developmental acquisition program), as well as other selected decisions as implemented according to policy to help round out the review (DoDI 5000.02, 2017; 10 U.S.C.; DAU, 2002, 2004, 2017b; Parker, 2011, MCSC, 2017; NAVAIR, 2015). As with the three slides that follow, the slide above—on requirements and technical, risk, and budgetary decisions—summarizes the results of this functional analysis. Key decisions as defined by statute and policy are listed in column 2 and are grouped by the larger domain (column 1) in which they fall. The types of analyses that support each decision are listed in column 3. The analytic capabilities that enable those supporting analyses are then listed in columns 4 through 7. They include the parts of the DoD AWF involved in the analysis, other resources or communities involved, a brief characterization of the analytic approaches, and the key data types that are generally available and used in the analyses. Finally, the last column identifies some gaps and challenges that the DoD faces in each analysis.

For example, market research is conducted to inform the decision on whether the technology to be used or further refined after Milestone B duplicates technology that could serve the program's needs and that exists in the broader market. That market research is conducted by the PM and by supporting engineers in both the program office and DoD laboratories, UARCs, and FFRDCs. Internet searches, technology assessments, and requirement analyses inform the market research. Key data are market information and technical literature. A challenge is access to PROPIN and other sensitive data for nongovernment analysts in the program office, laboratories, UARCs, and FFRDCs (especially because potential technology alternatives may be closely held by the owners of the intellectual property).

Acquisition decisionmaking requires a wide range of supporting analysis (Strategy and Trade-offs) *(Milestone B examples)*

			Analytic Capabilities				Gaps and Challenges
Domain	**Decision**	**Supporting Analyses**	**DoD AWF**	**Other Resources**	**Analytic Approaches**	**Key Data: Available**	
Strategy	Acquisition strategy	Acquisition strategizing	PM, contracting, FM, Leader-ship	Buying commands; FFRDCs	Acquisition strategizing; objectives; competition strategy; benefits analysis; market research; bundling; business strategy; contracting strategy (below); cooperative opportunities; general equipment valuation; Industrial-base analysis; intellectual property analysis and strategizing; architecture (e.g., modular open system?); multi-year procurement analysis; risk analysis; Small-business analysis (SBIR/STTR)	Requirements, system concepts, industrial base and competitive/cooperative opportunities, small-business capabilities and contributors, risks, cost estimates, equipment design and costs, intellectual property rights and options, architectural standards and options, procurement options	
Trade-offs	Trade-offs made and appropriate	Cost, schedule, and performance trade-off; Risk analysis; Framing assumptions; Budgetary trade-offs	PM	Service Chief; FFRDCs	Benefit/cost analysis; cost, schedule, and performance relationship modeling; risk analysis		
	Alternatives considered	AoAs; Market research	PM, Engr., S&T Manager	CAPE; DoD Labs and FFRDCs; Requirements	Operational military M&S; Technology forecasts; tech. maturity models; risk models	Operational systems, capabilities, missions, and intelligence; requirements; relating systems; Kill chains; Technology and market data; cost data	Analyst access to PROPIN.

SOURCES: DoDI 5000.02; 10 U.S.C.; DAU (2002, 2004, 2017b), Parker (2011), MCSC (2017); NAVAIR (2015); author expertise and experience

RAND CORPORATION

NOTE: AoA = analysis of alternatives; FM = financial management; S&T = science and technology; M&S = modeling and simulation.

This slide summarizes the results of our functional analysis for decisions related to strategy and tradeoffs (DoDI 5000.02, 2017; 10 U.S.C.; DAU, 2002, 2004, 2017b; Parker, 2011, MCSC, 2017; NAVAIR, 2015). For example, the milestone decision authority (MDA) needs to conclude whether tradeoffs between cost, schedule, performance, and risks have been made and whether they are appropriate. Analyses consist of trade-off analysis, risk analysis, and budgetary analysis. The service chief has the primary responsibility for making cost, schedule, and requirement trade-offs; the MDA, PEO, and, especially, the PM (supported by program staff and FFRDCs) make detailed programmatic analyses and assess trade-offs not illustrated in this slide. Specific analytic approaches include benefit-cost analysis; modeling the relationship among cost, schedule, and performance; and risk analysis.

Acquisition decisionmaking requires a wide range of supporting analysis (Cost)

(Milestone B examples)

			Analytic Capabilities				Gaps and Challenges
Domain	**Decision**	**Supporting Analyses**	**DoD AWF**	**Other Resources**	**Analytic Approaches**	**Key Data: Available**	
Cost	Costs estimated and reasonable	Cost analysis: developmental and market analysis	CE; Contracting; PM	SETA support contractors; UARCs; FFRDCs	Cost analysis (spreadsheets; custom tools; BI)	Prior and current system cost data	Analyst access to PROPIN and smaller program prior cost data.
	Cost within goals	Affordability analysis; Cost analysis: developmental and market analysis	PM	FFRDCs; UARCs			
	Funding available	Budgetary analysis (POM and FYDP)	FM; CE; PM; Lifecycle logistics	Resource planners (G8, N8, A8); FM; Leadership			
	Life-cycle sustainment planned and costed	Life-cycle sustainment planning and cost estimating	T&E, Lifecycle Logistics, FM, PM	Sustainment and logistics centers; FFRDCs	O&S cost estimating; reliability, material availability, operational availability, maintainability, supportability, corrosion, and survivability analysis (new technology and legacy systems)	Legacy O&S costs (e.g., VAMOSC); projected tested reliability and O&S costs of new system components	Estimates are uncertain (especially early in a program) due to projecting new system performance. Detailed sustainment data are not widely shared outside the Service. Externalities can greatly affect actual costs compared to estimates (e.g., changes in operational environments, OPTEMPO, physical and cyber battle damage, quantity, fuel costs, labor costs, healthcare costs, logistical system configuration and investments).

SOURCES: DoDI 5000.02; 10 U.S.C.; DAU (2002, 2004, 2017b), Parker (2011), MCSC (2017); NAVAIR (2015); author expertise and experience

RAND CORPORATION

NOTE: POM = program objective memorandum; FYDP = Future Years Defense Program; SETA = systems engineering and technical assistance; BI = business intelligence; O&S = operations and support; VAMOSC = Visibility and Management of Operation and Support Costs; OPTEMPO = operations tempo.

The slide above summarizes the results of our functional analysis related to cost (DoDI 5000.02, 2017; 10 U.S.C.; DAU, 2002, 2004, 2017b; Parker, 2011; MCSC, 2017; NAVAIR, 2015). For example, the MDA must decide whether life-cycle sustainment is well planned and costed. Those plans and estimates are made by PM, supported by the T&E, logistics, and financial management personnel and by sustainment and logistics centers and FFRDCs. Specific analytic approaches include O&S cost estimating and analysis of system reliability, material availability, operational availability, maintainability, supportability, corrosion, and survivability analysis (new technology and legacy systems). Data include legacy O&S data (e.g., in the VAMOSC system), projected and tested reliability, and O&S costs of new system elements. These estimates and plans are challenging because estimates are uncertain (especially early in a program) because of projecting new system performance. Also, detailed sustainment data are not widely shared outside the military departments. In addition, externalities can greatly affect *actual* costs (e.g., changes in operational environments, OPTEMPO, physical and cyber battle damage, quantity, fuel costs, labor costs, health care costs, and logistical system configuration and investments). Finally, O&S cost estimates might not receive as much attention as development and production estimates, since O&S costs are incurred so far in the future.

Acquisition decisionmaking requires a wide range of supporting analysis (Schedule, Compliance, Contracting, Production, Portfolio) *(Milestone B examples)*

			Analytic Capabilities				Gaps and Challenges
Domain	**Decision**	**Supporting Analyses**	**DoD AWF**	**Other Resources**	**Analytic Approaches**	**Key Data: Available**	
Schedule	Schedule estimated and reasonable	Schedule analysis; technical maturity and risk analysis	CE; Contracting; PM	FFRDCs			
Compliance	Program compliant with laws and policies	Legal and policy analysis	All but esp. PM	OGC; FFRDCs			
Contracting	Contracting strategy developed and appropriate	Contracting strategy and analysis	Contracting, FM, PM	Buying commands; FFRDCs	Contract-type determination; Termination liability estimating	Contract execution risks; contractor investments and costs	
Production	MS C, FRP, and production management	Engineering, manufacturing, T&E, quality, reliability, contracting, audits, and payments	PQM, T&E, Engr., Facilities Engr., IT, Logistics Contracting, Purchasing, Auditing, FM, PM,	Test centers, FFRDCs/UARCs	Numerous specialty-specific approaches	Designs, production approaches, quality, scrap rates, failure rates, reliability	Detailed data rarely available to higher-level oversight organizations
Portfolio	Program satisfies larger portfolio needs		Engr., PM; Leadership	Requirements, FFRDCs, Systems and Buying Commands	Ad hoc	Kill chains; JCAs; operational concepts; Program and system interdependencies and schedules; Current operational systems (commodity, inventory, age, condition, capabilities)	Portfolio analysis and data are largely ad hoc, depending on oversight workforce knowledge and unstructured analysis. Funding flexibility
	Program budget sufficient given larger portfolio interdependencies		Engr., FM, PM, Leadership	Requirements, FM, Systems and Buying Commands, FFRDCs	Ad hoc	Program and system interdependencies and schedules	Portfolio analysis and data are largely ad hoc by oversight workforce knowledge and unstructured analysis

RAND

SOURCES: DoDI 5000.02; 10 U.S.C.; DAU (2002, 2004, 2017b), Parker (2011), MCSC (2017); NAVAIR (2015); author expertise and experience

NOTE: MS C = Milestone C; FRP = full-rate production; OGC = Office of General Counsel; JCA = joint capability area.

The slide above summarizes the results of our functional analysis related to schedule, compliance, contracting, production, and portfolio at Milestone B (DoDI 5000.02, 2017; 10 U.S.C.; DAU, 2002, 2004, 2017b; Parker, 2011; MCSC, 2017; NAVAIR, 2015). It also adds entries for needed analysis related to portfolios. Portfolio analysis is less systematic and less mature than the other areas but important to note, since it can inform broader decisions related to budgets, requirements, and cross-program coordination, offering an integrated view on the net capabilities provided to operational DoD units. Currently, portfolio analysis is largely ad hoc, with elements performed at the PM, PEO, and oversight staff functions. Issues are raised, for example, during budget planning and reviews, but the DoD is currently exploring better ways to organize and conduct portfolio analysis.

Based on this functional analysis, we found a wide range of data analysis supporting major acquisition decisions

- Consider key elements of decisions at a program's Milestone B and selected other decision points
  - Extensive analysis is conducted to support acquisition decisions
    - *Some analysis is statutory, while other is policy and management driven*
  - Major analyses are conducted to support various functional domains
    - *e.g., requirements, technology, risk, budgets, strategy, tradeoffs, cost, schedule, compliance, contracting, production, and portfolios*
  - There are opportunities for improvement. For example:
    - *Portfolio analysis (e.g., program's relationship to the larger portfolio, mission, and kill-chains)*
    - *Data access and sharing*
    - *Better O&S metrics that separate externalities from acquisition-controlled aspects*
    - *Improved tracking between budgets and acquisition elements (e.g., programs, contracted services, procurements, and acquisition activity costs)*

As our examination illustrates, extensive DoD data analysis currently supports management decisions. Some analysis is statutory, whereas other analysis is driven by policy and management. These analyses cut across various functional domains in acquisition, such as requirements, technology, risk, budgets, strategy, trade-offs, cost, schedules, compliance, contracting, production, and portfolios. Despite the breadth of these analyses, clear opportunities for improvement exist, including expansion of portfolio analysis and improvements in data access and sharing to broader analytic communities.

Although we found extensive data analytics in place to support DoD acquisition decisions, problems or challenges still arise as to whether or not that information is used in decisionmaking regarding major acquisition programs. To better understand how well the data analytics outlined in the prior section support acquisition decisions and outcomes, we examined existing published analysis on 13 major acquisition programs that had known decisional challenges.

It is important to note the inherent limitations of such reviews. The sample does not necessarily reflect the broader set of acquisition programs. Also, some older programs are included, given that their postmortems are well documented, but they do not necessarily reflect the DoD's current data analytics situation. In addition, the situations in the programs discussed are very complex and detailed. We sought to extract the data analytics perspectives from these larger stories, but deeper discussions (which would better convey an understanding of the complications, analyses, and decisions in each case) are beyond the scope of this effort.

Literature review revealed that data analysis capabilities existed on many high-visibility acquisition programs, but challenges remained

- Examined 13 high-visibility acquisition programs
  - Milestone B dates from 2001–2016
  - All six programs with PARCA-identified root-causes where information was available but was not fully acted on by leadership, plus other historical examples (FCS, LCS)
  - Affordability cases
- Results (note: there is some overlap where programs have multiple "issues" and appear in the counts more than once)
  - 8 involved available data analysis overweighed by other decision factors
  - 6 involved data-analysis capability shortfalls
    - *1 affordability issue (now addressed by new process; may have been known at the time)*
    - *3 capacity shortfalls*
    - *4 with analytic issues*
  - 2 involved successes (new affordability process working)
  - 1 involved a quantity change
- Note:
  - OSD PARCA found that other considerations led to acquisition decisions that ran counter to available certain data and analytic results in 26 percent of cases they analyzed (6 of 23 root-cause analyses)
  - Samples do not reflect other programs that did not have major issues

NOTE: PARCA = Office of Performance Assessment and Root-Cause Analyses; FCS = Future Combat System; LCS = Littoral Combat Ship.

We examined 13 programs with milestone B dates ranging as far back as 2001. This set included all six programs for which OSD's independent PARCA identified a root cause when information was available but was not fully acted on by leadership (see USD[AT&L], 2016a, p. 28). We also sampled programs that had Nunn-McCurdy cost breaches or affordability issues (affordability and affordability analysis are a good example of a challenging acquisition decision). Of course, these are only some examples of problematic programs, and they do not get into the types of execution issues that program managers deal with on a daily basis. PARCA's analyses focus on root causes of primary problems rather than other challenges.

In our review of these programs, we identified two general types of issues related to data analytics: those related to decisional use of the available data analytics and those related to a capability shortfall in data analytics. The former includes eight instances in which information was available but was not fully acted on by leadership (Pernin et al., 2012; Work, 2013; USD[AT&L], 2016a). The latter consists of six cases involving some type of data analysis capability shortfall (e.g., an affordability issue, capacity shortfalls, or issues with the analysis).

In addition to cases in which strategic reasons led to decisions that run counter to certain available data analytics results, we note that 26 percent (six of 23) independent root-cause analyses (breach and discretionary) performed by PARCA involved decisionmaker "failure to act on information" (USD[AT&L], 2016a; Director of PARCA, 2017, 2018). Thus, although this is a concern, it has been limited to about one-quarter of problematic programs since 2010 and thus a smaller percentage of the larger set that includes programs without major issues.

Of the 13 programs we examined, two demonstrated successful results from the new affordability process. One program involved a quantity change and thus was not related to analysis (or related to acquisition performance).

These insights show that, in terms of these programs with major decision-related issues, data analytics are often available (i.e., the primary cause is not a general lack of data analytics capability). Sometimes, other considerations led to acquisition decisions that ran counter to available data analytics results. Also, analytic functions are not perfectly executed or always sufficiently resourced, but this level of review does not appear to show a major shortcoming. In addition, old programs, such as Global Hawk, F-35, FCS, Integrated Defensive Electronic Countermeasures, Block 2/3 (IDECM B2/B3), and *Zumwalt*-class guided missile destroyer (DDG-1000) are not necessarily reflective of current data analytics conditions in the DoD but are included because their causes are well documented and they help illustrate the types of possible concerns. Thus, they should not be used to critique the DoD's current situation.

On challenging programs, data analytics are often available but may be imperfectly utilized or executed

Data Analysis Issue?			#	Milestone B or III Date (FY)															
Decision	Capability			2001	2002	2003	2004	2005	2006	2007	2008	2009	2010	2011	2012	2013	2014	2015	2016
		Issues:																	
Yes	No	Other strategic consideration affected decisions that ran counter to available data and analytic results	8	Global Hawk; F-35		FCS	Navy ERP			GCSS-MC				LCS	OCX; AF ECSS*				
No	Yes	Affordability process not yet in place						DDG 1000											
No	Yes (capacity)	Analytic workforce insufficient on program: staffing or expertise	3			FCS		AAG							OCX				
No	Yes	Analytic issues: design errors, inadequate understanding, or limited awareness	4	F-35			Navy ERP	AAG		GCSS-MC									
No	No	Schedule-driven risks													OCX				
No	No	Other		Global Hawk		FCS									OCX				
		Successes:																	
No	No	Affordability data informed actions that drove cost reductions													JLTV				SSBN 826
		Non-Issue:																	
No	No	Quantity change				IDECM B2/B3													

SOURCES: Director of PARCA (2010a, 2010b, 2011a, 2011b, 2011c, 2013, 2016, 2017, 2018), Aronin et al. (2011), Blickstein et al. (2011, 2012), Pernin et al. (2012), Work (2013), USD(AT&L) (2016a).

NOTE: ERP = enterprise resource planning; AAG = Advanced Arresting Gear; GCSS-MC = Global Combat Support Systems—Marine Corps; OCX = Operational Control System; AF = Air Force; ECSS = Expeditionary Combat Support System; JLTV = Joint Light Tactical Vehicle.

The previous slide maps the 13 case programs against specific issues, successes, or nonacquisition issues. Here, programs are shown by the year of their Milestone B approval—except the Air Force's ECSS, which did not have a Milestone B and so is shown in the year it was canceled (see Aronin et al., 2011). Note that programs might have multiple concerns. For most of these programs, we used the PARCA root-cause analysis as the basis for our review, but we used our personal insight and available literature for FCS, LCS, JLTV, and the SSBN 826 programs. PARCA root-cause analyses are reported by the Director of PARCA (2010a, 2010b, 2011a, 2011b, 2013, 2016, 2017, 2018) and also summarized in USD(AT&L) (2016a). Other sources we analyzed were Pernin et al. (2012) and Work (2013).

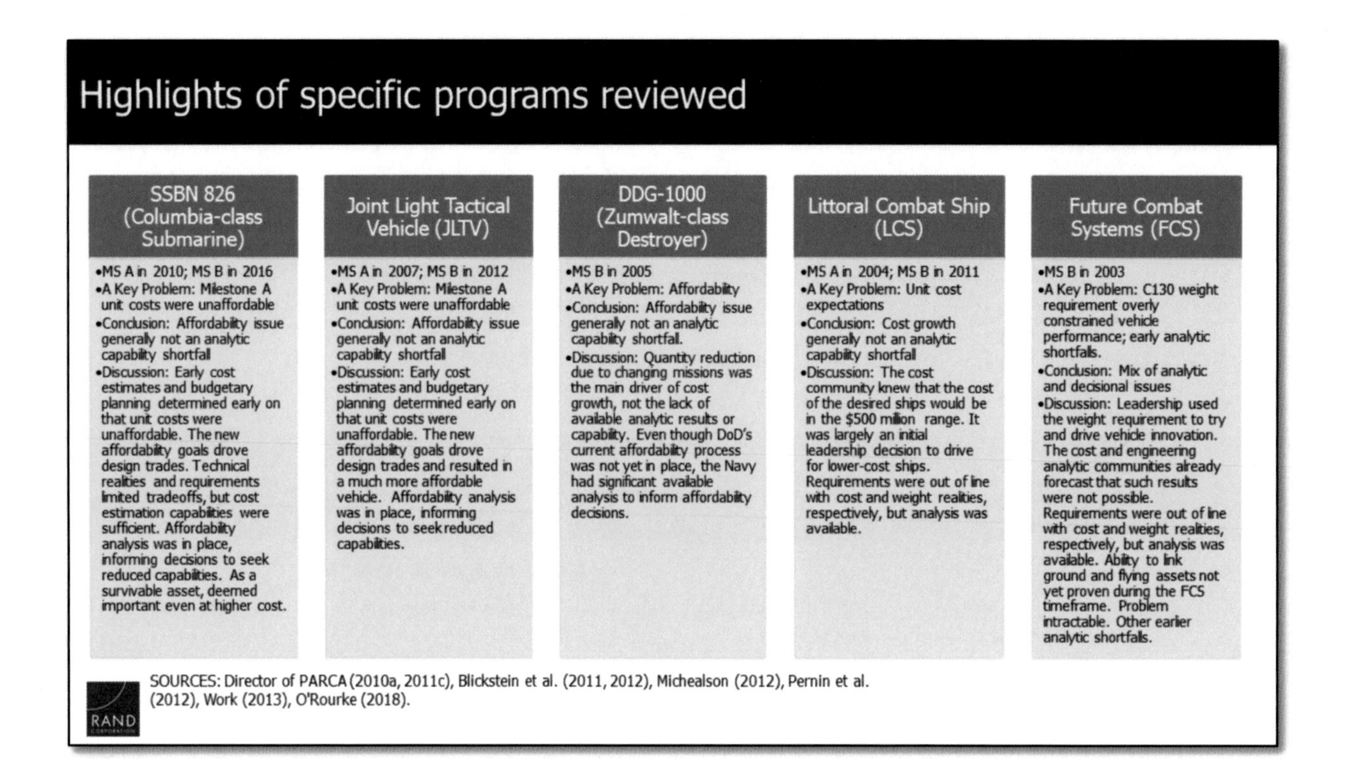

Highlights of Programs Reviewed

The following descriptions summarize, from a data analytics perspective, our review of the non-ERP programs. Although these cases are complicated, we extracted from the existing literature a key problem that relates to the use of data analytics (or any lack thereof). Thus, these brief highlights should not be taken as comprehensive reviews but rather as insights into the question of the use of data analytics for program decisions.

SSBN 826 (*Columbia*-class ballistic missile submarine, nuclear-powered)—Milestone A in 2010 and Milestone B in 2016:

- *A key problem:* Milestone A unit costs were unaffordable.
- *Conclusion:* The affordability issue was generally not an analytic capability shortfall.
- *Discussion:* Early cost estimates and budgetary planning determined that unit costs were unaffordable. The new affordability goals drove design trades. Technical realities and requirements limited trade-offs, but cost-estimation capabilities were sufficient.

Affordability analysis was in place, informing decisions to seek reduced capabilities. As a survivable asset, it was deemed important even at higher cost. See Michealson (2012, p. 6).

JLTV—Milestone A in 2007 and Milestone B in 2012:

- *A key problem:* Milestone A unit costs were unaffordable.
- *Conclusion:* Affordability issue generally not an analytic capability shortfall
- *Discussion:* Early cost estimates and budgetary planning determined that unit costs were unaffordable. The new affordability goals drove design trades and resulted in a much more affordable vehicle. Affordability analysis was in place, informing decisions to seek reduced capabilities. See Michealson (2012, pp. 14–15).

DDG-1000 (*Zumwalt*-class guided missile destroyer)—Milestone B in 2005:

- *A key problem:* Affordability was a potential problem throughout.
- *Conclusion*: The affordability issue was generally not an analytic capability shortfall
- *Discussion*: Even though the DoD's current affordability process was not yet in place, the Navy did affordability analysis. Affordability was known to be a potential problem. Ultimately, the quantity was reduced, and alternatives were pursued. The land attack mission also came into question as adversary capability increased from the shore. See Director of PARCA (2010a, 2011c), Blickstein et al. (2011, 2012), and O'Rourke (2018).

LCS—Milestone A in 2004 and Milestone B in 2011:

- *A key problem:* Unit cost expectations were a concern.
- *Conclusion:* Cost growth was generally not an analytic capability shortfall.
- *Discussion:* The cost community knew that the cost of the desired ships would be in the $500 million range. It was largely an initial leadership decision to drive for lower-cost ships. Requirements were out of line with cost and weight realities, but analysis was available. See, for example, Work (2013).

FCS—Milestone B in 2003:

- *A key problem:* The C-130 weight requirement overly constrained vehicle performance, and there were early analytic shortfalls.
- *Conclusion:* There was a mix of analytic and decisional issues.
- *Discussion:* Leadership used the weight requirement to try to drive vehicle innovation. The cost and engineering analytic communities already forecast that such results were not possible. Requirements were out of line with cost and weight realities, but analysis was available. The ability to link ground and flying assets had not yet been proven during the FCS time frame. Thus, the technical problem was intractable. There were also analytic shortfalls in the design of the system concepts. See, for example, Pernin et al. (2012).

Highlights of specific programs reviewed (concluded)

GPS Next-Generation Operational Control System (OCX)

- MS B in 2012
- A Key Problem: Cost growth
- Conclusion: Cost breach mostly a schedule risk problem, but workforce shortfalls contributed
- Discussion: Three root causes: unrealistic schedule driven by the need to sustain the GPS constellation; inexperience driving underestimation of the cost to fully implement new Information Assurance instructions; and systems engineering (insufficient software engineering expertise and tenure in the program office; quality FFRDC analysis but not sufficiently listened to; insufficient contractor requirements decomposition and configuration management).

Integrated Defensive Electronic Countermeasures (IDECM) Block 2/3 Subprogram

- MS III in 2003
- A Key Problem: Quantity change
- Conclusion: Cost breach generally not an analytic capability shortfall
- Discussion: Quantity reduction of ~69% caused the breach.

Advanced Arresting Gear (AAG)

- MS B in 2005
- A Key Problem: RDT&E cost growth from engineering design issues
- Conclusion: Analytic capability shortfall in engineering workforce levels and expertise.
- Discussion: Design issues were addressed late due to insufficient government and contractor engineering workforce and program manager attention to design issues.

RQ-4A/B Global Hawk

- MS B in 2001
- A Key Problem: Cost growth
- Conclusion: Continued underestimation may have been an analytic shortfall.
- Discussion: Available information included: known but unfunded requirements due to budgetary pressures; did not change spiral strategy of deferring requirements; reliability program was directed but not pursued. Other issues included: (a) an unrealistic schedule based upon the continued underestimation of the differences between aircraft types and systems integration requirements for the payloads; and (b) additional requirements and changes in aircraft type mixes.

F-35 Joint Strike Fighter

- MS B in 2001
- A Key Problem: Cost growth
- Conclusion: A mix of poor analytic execution, insufficient analytic capability, and available information not fully acted upon;
- Discussion: Two root causes: (a) flawed programmatic, estimating, and technological assumptions at inception, and (b) poor systems engineering, ineffective contractual incentives, and an environment in which there was a general reluctance to accept unfavorable information.

SOURCES: Director of PARCA (2010b, 2011b, 2016, 2017, 2018), USD(AT&L) (2016a).

RAND

NOTE: MS = milestone.

GPS Next-Generation OCX—Milestone B in 2012:

- *A key problem:* Cost growth was a problem.
- *Conclusion:* Cost breach was mostly a schedule risk problem, but workforce shortfalls contributed.
- *Discussion:* Three root causes were identified: unrealistic schedule driven by the need to sustain the GPS constellation, inexperience driving underestimation of the cost to fully implement new information assurance instructions, and systems engineering (insufficient software engineering expertise and tenure in the program office, quality FFRDC analysis that was not sufficiently listened to, and insufficient contractor requirements regarding decomposition and configuration management). See Director of PARCA (2016) and USD(AT&L) (2016a).

IDECM B2/B3 Subprogram—Milestone III in 2003:

- *A key problem*: Quantity change caused a Nunn-McCurdy breach.
- *Conclusion*: Cost breach was generally not an analytic capability shortfall.
- *Discussion*: Quantity reduction of approximately 69 percent caused the breach. See Director of PARCA (2018).

AAG—Milestone B in 2005:

- *A key problem:* RDT&E cost growth resulted from engineering design issues.

- *Conclusion:* There was an analytic capability shortfall in engineering workforce levels and expertise.
- *Discussion:* Design issues were addressed late because of insufficient government and contractor engineering workforce and program manager attention to design issues. See Director of PARCA (2017).

RQ-4A/B Global Hawk—Milestone B in 2001:

- *A key problem:* Cost growth led to a Nunn-McCurdy breach.
- *Conclusion:* Continued underestimation might have been an analytic shortfall.
- *Discussion:* There was information that was available but not fully acted on by leadership: there were known but unfunded requirements, because of budgetary pressures; the spiral strategy of deferring requirements was not changed; and a reliability program was directed but not pursued. Other issues were (1) an unrealistic schedule based on the continued underestimation of the differences between aircraft types and system integration requirements for the payloads and (2) additional requirements and changes in aircraft type mixes. See Director of PARCA (2011b).

F-35 Joint Strike Fighter. MS B in 2001.

- *A key problem*: Cost growth led to a Nunn-McCurdy breach.
- *Conclusion*: There was a mix of poor analytic execution, insufficient analytic capability, and information that was available but was not fully acted on by leadership.
- *Discussion*: Two root causes were identified: (1) flawed programmatic, estimating, and technological assumptions at inception and (2) poor systems engineering, ineffective contractual incentives, and an environment in which there was a general reluctance to accept unfavorable information. See Director of PARCA (2010b).

The motivations that lead to decisions that run counter to available data analytics results can inform possible mitigations

Example motivations and issues:

- Stretch goal
  - *Issue:* In retrospect may have been unrealistic in time, technology, or physics
- Tried to fix problems
  - *Issue:* Problems were more fundamental than believed
- Satisfy external pressures (e.g., schedule, technical capabilities, budget)
  - *Issue:* May lead to unrealistic acquisition demands
- Must judge risks
  - *Issue:* Decisions are inherently subjective (that is why we have decision-makers)

Possible mitigations:

- Ensure that DoD leadership have balanced decision incentives and authorities
  - Example: OSD and Joint Staff oversight has more independence to raise issues and broader equities
- Training for future leaders to understand strengths and limits of data analytics (and technology)

Although it is beyond the scope of this study to dive deeply into why other considerations led to acquisition decisions that ran counter to available data analytics results, it is useful to give some examples from these cases. Caution is warranted, since, in hindsight, certain problems may be more obvious now than when the decisions were made; however, in some cases, experts knew that the decision was flawed at the time.

Some leaders have not fully utilized available analysis to push for a stretch goal and motivate a breakthrough or new thinking. For example, FCS was given a nontradable requirement that the vehicle be light enough to be air lifted by a C-130 (e.g., see Pernin et al., 2012). The intent was to motivate technologists and designers to devise a solution. Unfortunately, this stretch goal was too aggressive and was held for too long without leadership changing course, despite an analysis on the maturity of current technology. Perhaps decades of research and breakthroughs might yield a solution, but that expectation turned out to be unrealistic for the time. Also, some decisions are motivated by a desire to try to fix a problem program even though the data show that the problems might not have a solution.

In addition to the output of data analytics, decisionmakers need to take into account external pressures that may exist, such as meeting a warfighter's schedule demands or needed technical capabilities or trying to acquire needed capabilities despite limited budget capabilities. Balancing these trade-offs can lead to or further exacerbate acquisition problems, such as cost growth, schedule slippage, and poor technical performance.

Furthermore, some decisions inherently involve risks. The decisions to assume certain risks may frustrate those raising the risks (or those critiquing the decision in hindsight), but what may appear to be disregard of data analytics results could be risk judgments that might not align with a different person's assessment.

One idea for mitigating these issues is to seek better ways to balance DoD leadership incentives and authorities. We realize that competing incentives cannot be eliminated, but checks and balances, as well as independent voices, can help. For example, although program oversight by OSD and the Joint Staff can slow programs and introduce bureaucratic burden, they also introduce independence to raise issues and seek broader balance and trade-offs. Another idea is to continue introducing training for future leaders to understand strengths and limits of data analytics (and technology) and how to balance analysis with external constraints. Efforts at DAU and elsewhere (discussed later) could assist future acquisition leaders by providing a better understanding and appreciation of the information provided to them by data analytics.

What's in this Briefing

- Background
- Approach and methodology
- Findings on congressional questions
  - Q1–Q2: Extent of, and potential to increase, DoD data analytics capabilities
    - *Map decisions to analyses and data; identify gaps.*
    - *Data analytics availability and use on selected programs*
    - *DoD's information systems that enable data analytics*
    - *DoD's status against maturity models*
    - *Spending on the analytic workforce*
    - *Spending on IT data and analysis systems*
    - *Example limits in data availability and analytic challenges from this study*
  - Q3: R&D funding for data analytics capabilities
  - Q4: Private-sector best practices for efficient data collection and delivery
  - Q5: Anonymized data sharing with researchers and analysts
  - Q6: Appropriate data analytics and methods courses at defense training institutions
- Conclusions

The DoD's Information Systems That Enable Data Analytics

Questions 1 and 2 ask about "the extent to which data analytics capabilities have been implemented within the military services, DOD laboratories, test centers, and [FFRDCs] to provide technical support for acquisition program management." One way to gain an understanding of the current state in the DoD is through an analysis of the types of capabilities present in the DoD's centralized acquisition information systems. The reason for this analysis is that these information systems are utilized by a large portion of the workforce for a variety of purposes, including analysis. These information systems provide data and tools for the workforce and therefore enable the workforce to conduct analytics on acquisition data.

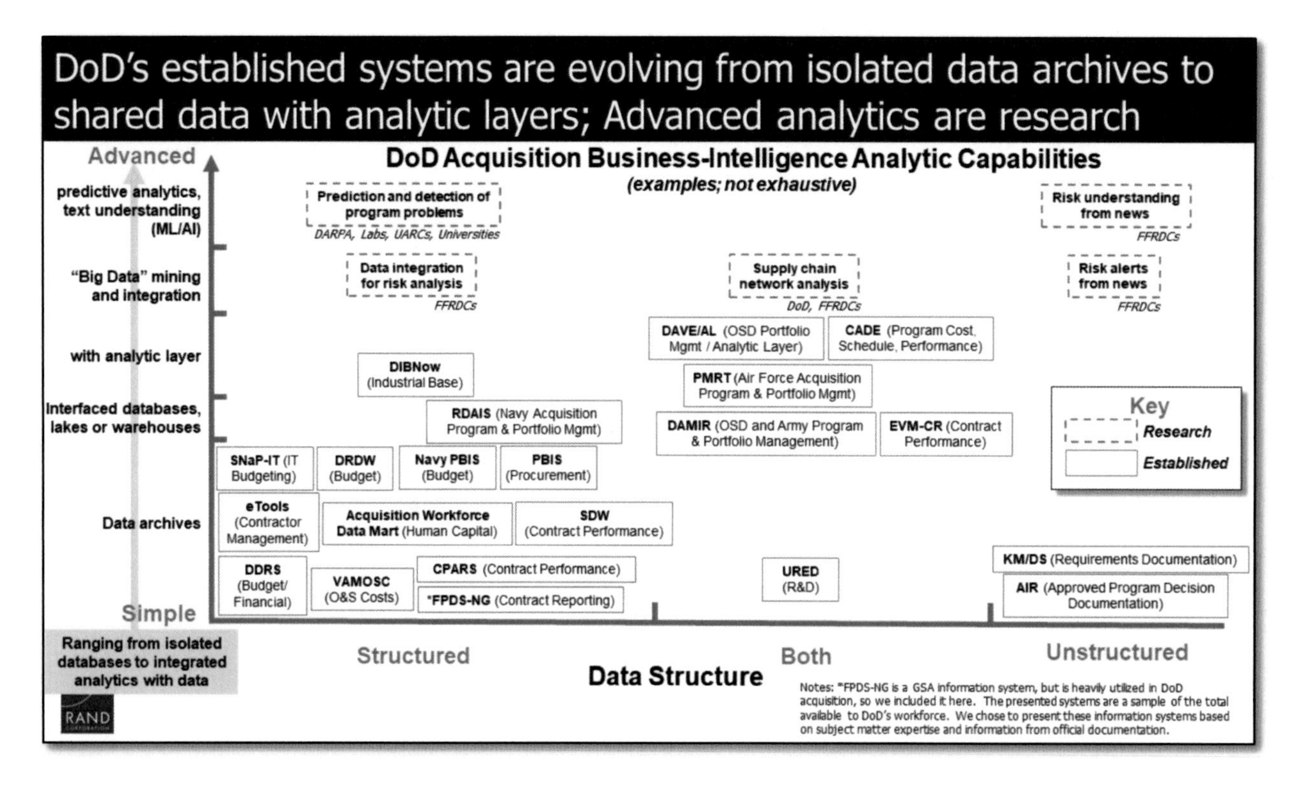

NOTE: AIR = Acquisition Information Repository; AL = analytic layer; CADE = Cost Assessment Data Enterprise; CPARS = Contractor Performance Assessment Reporting System; DAMIR = Defense Acquisition Management Information Retrieval; DARPA = Defense Advanced Research Projects Agency; DAVE = Defense Acquisition Visibility Environment; DIBNow = Defense Industrial Base Now; DDRS = Defense Department Reporting System; DRDW = DoD Resources Data Warehouse; eTools = Electronic Tools (Defense Contract Management Agency [DCMA]); EVM-CR = Earned-Value Management—Central Repository; FPDS-NG = Federal Procurement Data System—Next Generation; ML/AI = machine learning/artificial intelligence; PBIS = Procurement Business Intelligence Service (used by OSD and the Army) *and* Program Budget Information System (used by the Navy); PMRT = Project Management Resource Tools; RDAIS = Research, Development, and Acquisition Information System; SDW = Shared Data Warehouse; SNaP-IT = Select and Native Programming—Information Technology; URED = Unified Research and Engineering Database.

Acquisition data lay a foundation for the DoD's decisionmaking, execution, management, insight, and oversight of the weapon system acquisition portfolio. Acquisition data help inform, monitor, and achieve many objectives (e.g., promoting transparency in spending, conducting analyses for improved decisionmaking, and archiving decisions). Acquisition data are both *structured* (that is, immediately identified within an electronic structure, such as a relational database) and *unstructured* (that is, not in fixed locations but in free-form text). Such data are collected for a variety of statutory and regulatory requirements and at all levels, from program offices in the military departments to OSD, in both centralized and decentralized locations. The DoD also uses data from other federal information systems outside the department. The data themselves differ in time frame (with some dating back more than five decades), while the information systems that contain these data have different hardware, software, and interfaces. Technological improvements have helped the DoD improve data collection efficiency, quality, aggregation, ease of access and use, archiving, and analysis, among other characteristics; however, many information systems are difficult for users to navigate effectively and can take years to fully understand. Most systems are built for reporting, not analysis (McKernan et al., 2017).

The DoD has established many centralized information systems over the past 20 years in a variety of functional business areas (including R&D, requirements, budgeting, contracting, program cost, human capital, and acquisition oversight) to make data more available for management, analysis, and decisionmaking. Acquisition decisionmakers use these data to answer a wide variety of questions related to defense acquisition. Whereas the data sets vary considerably, many share two strengths: standardization and collection of acquisition-related information in one place where data can be input, accessed, and analyzed by those needing it. The DoD is very large, and acquisitions are accomplished by many different organizations. A centralized system with consistent formats helps improve data consistency and quality. In addition, analysis tools are layered on top of the data portions of some of the IT system architecture with varying levels of maturity (McKernan et al., 2017). These analytic tools surpass dashboards by enabling both custom and predefined analyses.

The previous slide provides a simplistic visualization of examples of various centralized information systems. The x-axis shows whether the data in those information systems are structured, unstructured, or both. The y-axis illustrates the capabilities of those information systems, ranging from simple storage to predictive analytics. This assessment is not meant to be a judgment of whether the DoD's capabilities are sufficient. Rather, it illustrates the range of capabilities across acquisition functions. The figure also illustrates that the DoD has established a variety of information systems over the past several years and matured prior capabilities in the military departments and OSD to provide data and more-advanced commercial business intelligence analytic software tools (e.g., Qlik Sense, Tableau). For example, OSD has moved to DAVE for acquisition program information; DAVE contains a recently added analytic layer for advanced data scientists. Cost analysts in OSD have also matured their analytic capabilities with CADE over the past several years. The Air Force has moved toward an advanced business intelligence capability leveraging Qlik Sense for program data with PMRT, whereas the Army and the Navy are both considering options to improve the analytic capabilities for their AWFs. The industrial base community has likewise added more-advanced capabilities with DIBNow. Other efforts that include big-data mining and integration and predictive analytics are being pursued through work in DARPA, DoD Laboratories, FFRDCs, UARCs, and universities (according to personal communications we had with DoD officials). These practices have created an environment more conducive to analytics; however, some obstacles still impede the DoD's continuing improvement in acquisition analytics. For example, DoD acquisition data management continues to be stovepiped by components and functional communities. Few formal (i.e., established) mechanisms exist to deal with cross-functional issues. This stovepiping has also resulted in varying levels of maturity in data management practices across the acquisition community, making it even more difficult to accomplish such tasks as sharing between major information systems.

Some advanced analytics are being pursued outside of DoD's centralized information systems, but more is possible

Example current research activities

- Text understanding and extraction of corporate news feeds
- Early detection and categorization of program problems
- Data integration for risk analysis
- Supply chain network analysis

Other potential research opportunities

- "Big Data" analysis
  - Large vehicle sensor data
  - Sustainment data trend and cause analysis
- Understanding and alerting on corporate problems in news and other data feeds

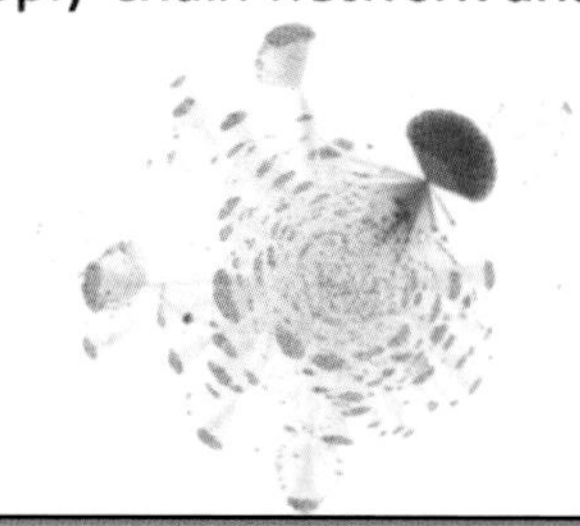

DoD is also pursuing advanced analytics, big data, and artificial intelligence / machine learning for system design, production, and within acquired systems, but this briefing focused more on data analytics for acquisition management rather than within acquired systems.

As noted, the DoD provides business intelligence analytic capabilities through multiple acquisition data information systems. Beyond these systems, the DoD has not invested a lot in capabilities for advanced analytics in its information systems; however, the DoD has some additional ongoing research activities that illustrate how advanced analytics are being pursued outside the DoD's centralized information systems. Some examples of these research activities include text understanding and extraction of corporate news feeds; early detection and categorization of program problems; data integration for risk analysis; and supply-chain network analysis. We are aware, for example, of ongoing unpublished research activities at FFRDCs at RAND and the Institute for Defense Analyses, and internet characterizations of these activities indicate that such research is also ongoing at other FFRDCs, laboratories, and research centers. However, data are not readily available to easily quantify the extent of these efforts.

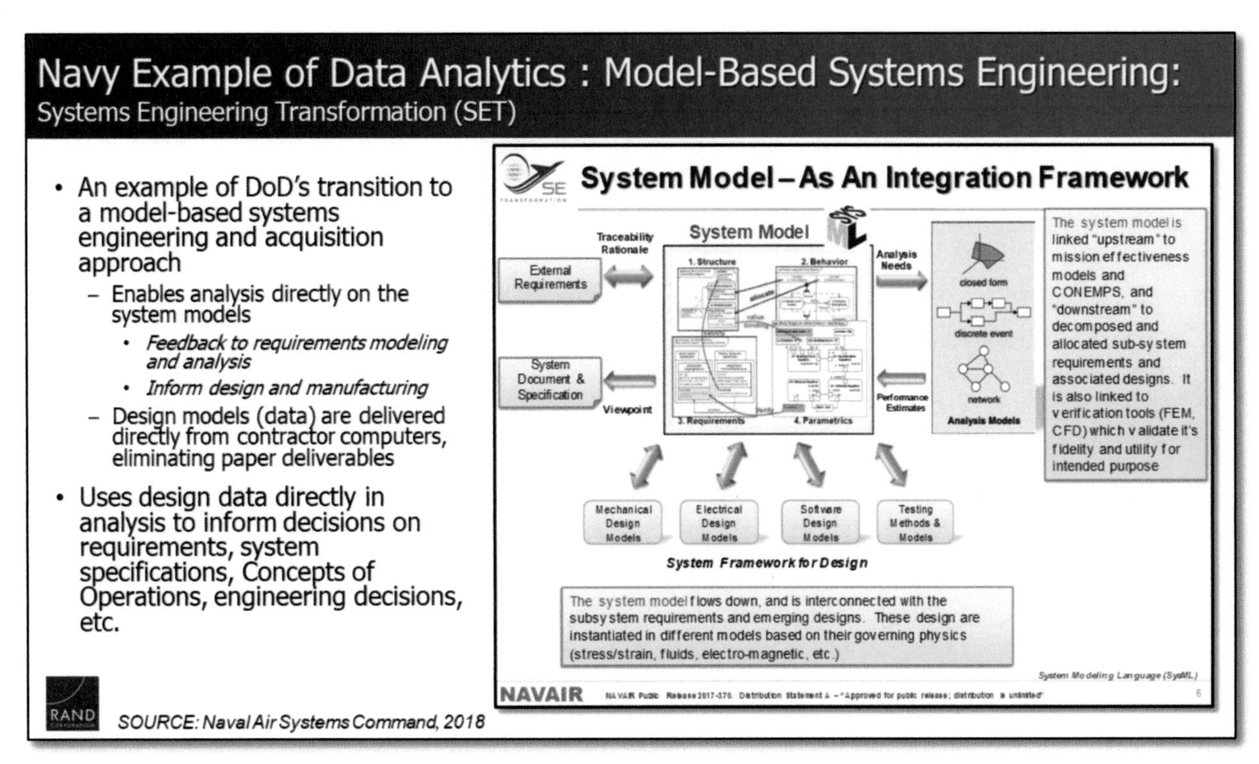

NOTE: SET = Systems Engineering Transformation; CONEMPS = concepts of employment.

Examples from the Navy and Air Force

To further illustrate the kind of data analytics capabilities in DoD components, we provide examples from the departments of the Navy and the Air Force.

Navy Example: Model-Based Systems Engineering

One example is from the NAVAIR's SET effort (NAVAIR, 2018). SET is one of the DoD's efforts to advance systems engineering and analysis from a diagram-based design and systems engineering to one in which a model of the system's various elements (mechanical, electrical, software) represents both the design and how it should function internally and with external systems. This advance allows engineering analysis tools to test the system's functions and how they are related to requirements. Results provide feedback to requirement modeling, analysis, and decisionmaking, as well as inform design and manufacturing going forward. Design models are delivered directly from the contractor's design computers, streamlining data delivery and eliminating paper deliverables.

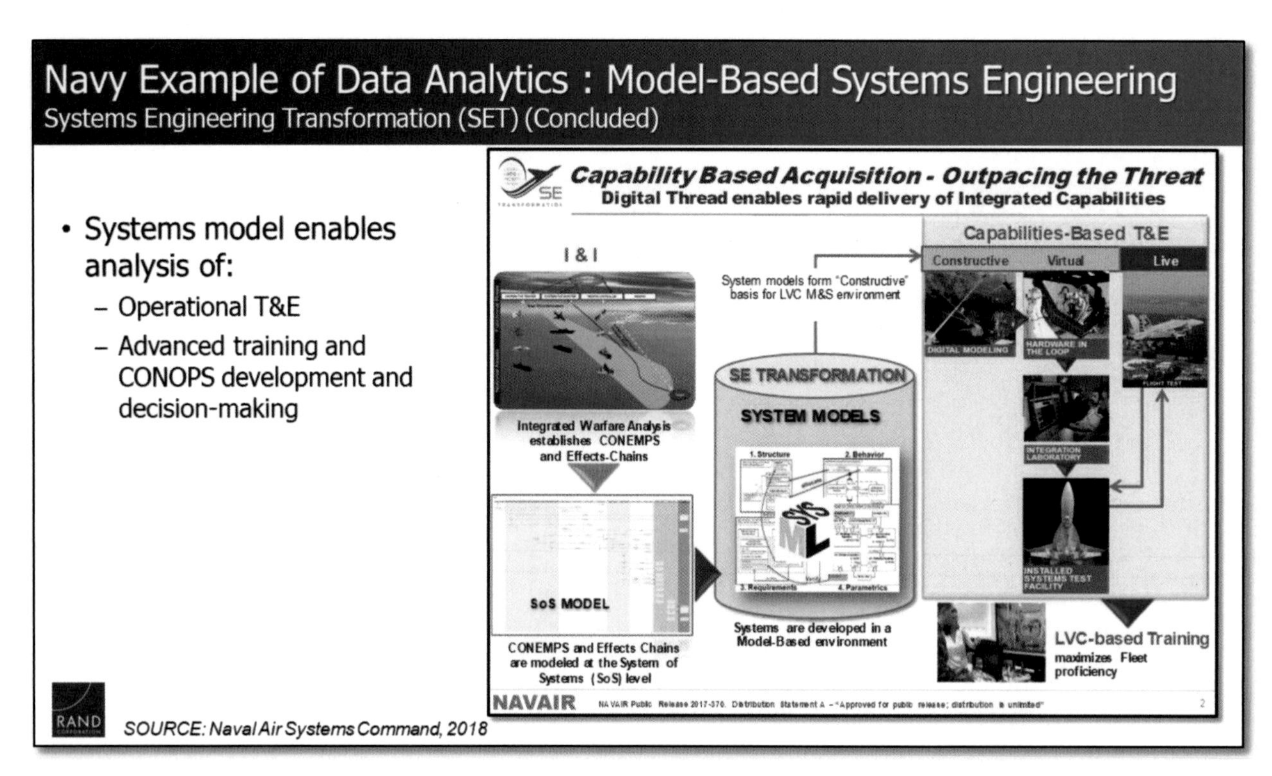

NOTE : LVC = live-virtual-constructive; CONOPS = concept of operations; SoS = systems of systems; SE = systems engineering.

In addition to enabling the delivery of requirements from requirements analysis systems, model-based systems engineering can form the basis of other analysis, such as M&S in support of operational T&E, as well as early training and CONOPS.

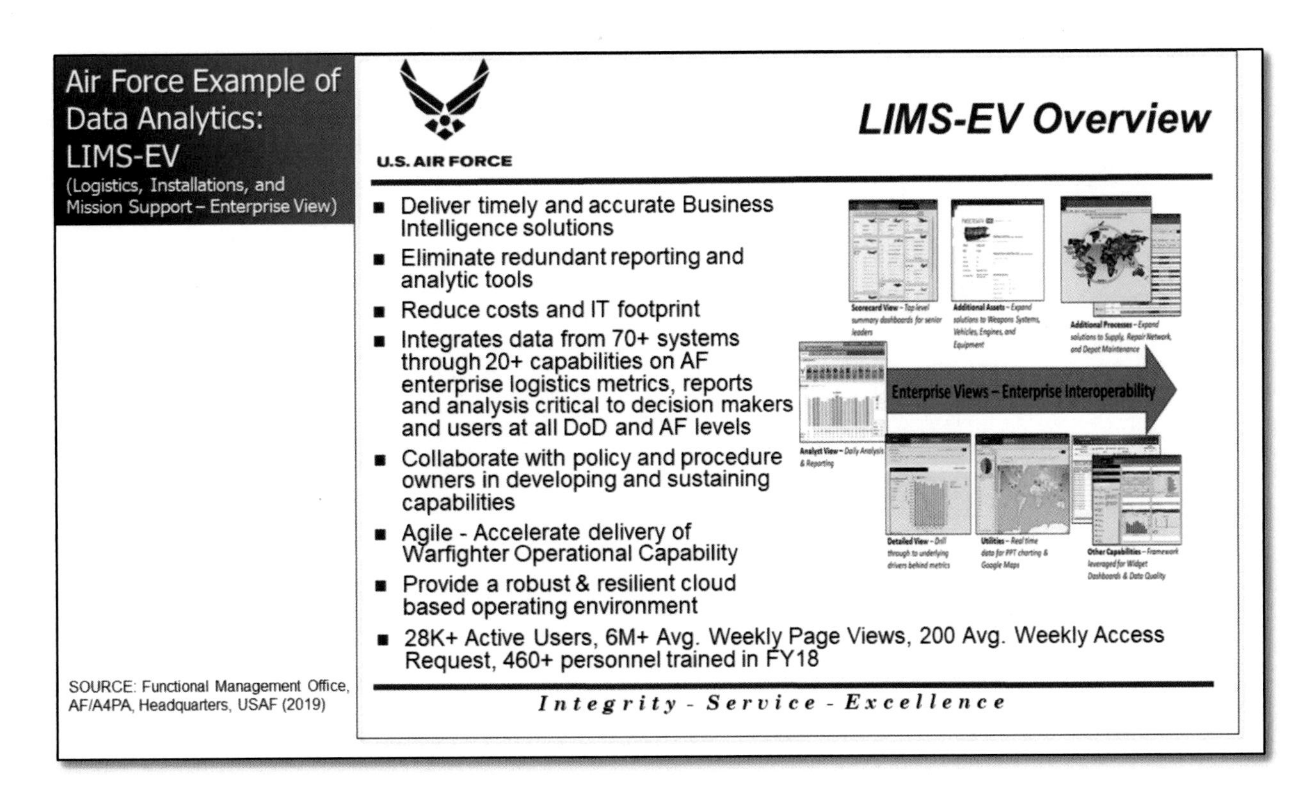

SOURCE: Functional Management Office, Logistics, Engineering and Force Protection, Resource Integration, 2019.
NOTE: LIMS-EV = Logistics, Installations, and Mission Support—Enterprise View.

Air Force Example: Integrated Logistics, Installations, and Mission Support Decision Analysis

The LIMS-EV system is an example of an integrated data analytics information system in the Air Force. The Air Force says that "LIMS-EV is an integrated suite of capabilities that serve as the overarching gateway to A4 enterprise Business Intelligence reporting and analytics" (Functional Management Office, Logistics, Engineering and Force Protection, Resource Integration, 2019). LIMS-EV integrates data from more than 70 systems, covering more than 40 capabilities on Air Force "enterprise logistics metrics, reports, and analysis critical to decision makers and users at all [Air Force] levels. . . . LIMS-EV is comprised of a host of capabilities spanning across Executive (leadership), Logistics Readiness, Requirements, Maintenance Repair and Overhaul, and Installation and Mission Support" (Functional Management Office, Logistics, Engineering and Force Protection, Resource Integration, 2019). LIMS-EV is accessed through the "USAF Portal" under the applications menu. AF/A4PA is the Air Force's LIMS-EV Functional Management Office, and the program office is Air Force Life Cycle Management Center/HNII.

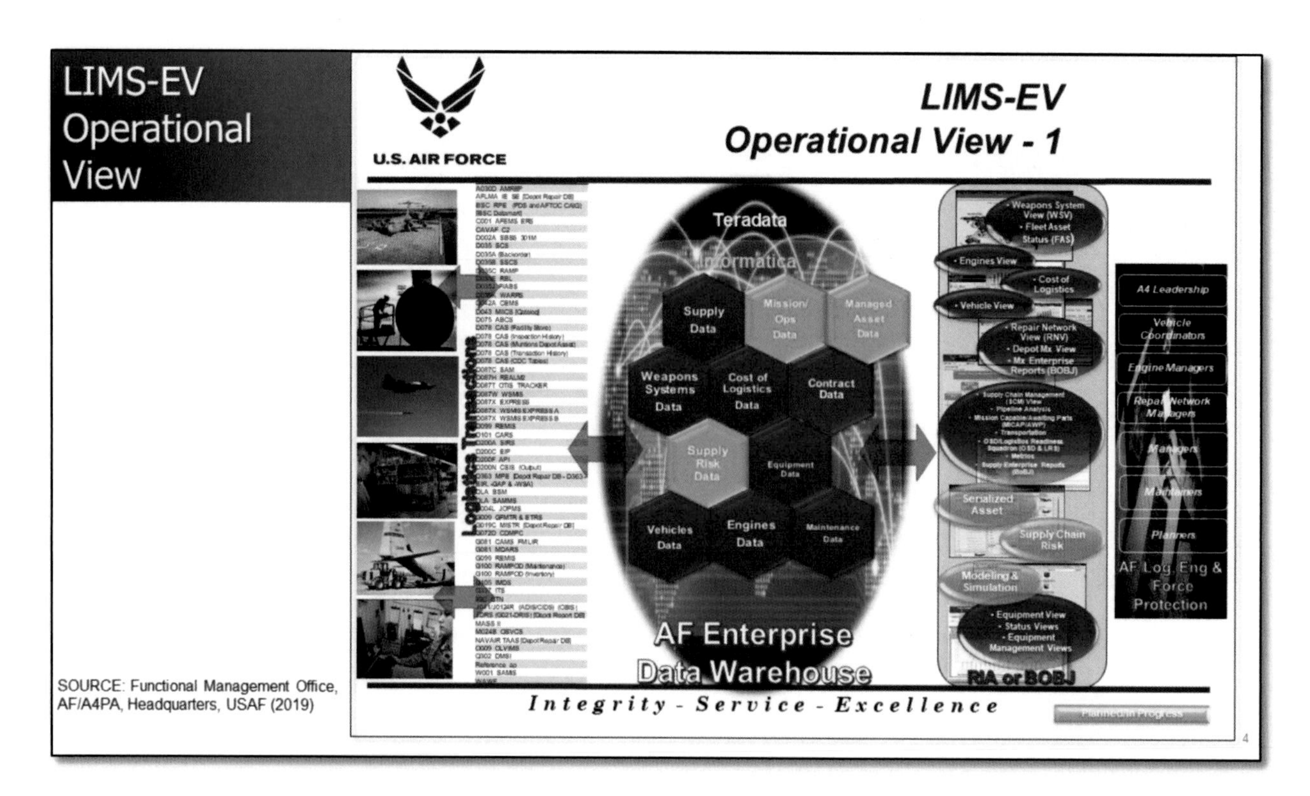

SOURCE: Functional Management Office, Logistics, Engineering and Force Protection, Resource Integration, 2019.

This operational view of LIMS-EV shows that it extracts logistics, maintenance, financial, operational use, and engineering data to enable various views and analysis for different functions, such as acquisition and sustainment engineers, managers, maintainers, planners, and leadership (Functional Management Office, Logistics, Engineering and Force Protection, Resource Integration, 2019). Again, it is beyond the scope of our effort to examine such systems in depth, but the example in the slide above helps illustrate that the DoD has moved toward better data integration and analysis. Investments and further development are continuing (e.g., budget exhibits show that RDT&E for LIMS-EV was $9,428,000 in FY 2017).

DoD's Status Against Maturity Models

What's in this Briefing

- Background
- Approach and methodology
- Findings on congressional questions
  - Q1–Q2: Extent of, and potential to increase, DoD data analytics capabilities
    - *Map decisions to analyses and data; identify gaps*
    - *Data analytics availability and use on selected programs*
    - *DoD's information systems that enable data analytics*
    - *DoD's status against maturity models*
    - *Spending on the analytic workforce*
    - *Spending on IT data and analysis systems*
    - *Example limits in data availability and analytic challenges from this study*
  - Q3: R&D funding for data analytics capabilities
  - Q4: Private-sector best practices for efficient data collection and delivery
  - Q5: Anonymized data sharing with researchers and analysts
  - Q6: Appropriate data analytics and methods courses at defense training institutions
- Conclusions

As part of our analysis to answer questions 1 and 2, we used private-sector maturity models to characterize the current environment in the DoD for acquisition analytics.

Maturity models provide an indication of how well an organization is accomplishing data-related tasks or processes

- Maturity models "...provide a way for organizations to approach problems and challenges in a structured way by providing both a benchmark against which to assess capabilities and a roadmap for improving them" (Caralli, Knight, and Montgomery, 2012, p. 2).
  - Maturity models use a qualitative or mixed-methods structured approach
- Reviewed four models that help identify best practices and provide a structure to compare DoD with the commercial sector
  - Gartner's Maturity Model for Data and Analytics
  - IBM's Maturity Model for Big Data and Analytics
  - U.S. Government's Federal Data Strategy "draft practices"
  - The Data Cabinet's Government-wide Data Maturity Model (DMM)
- Gartner and IBM focus on private sector practices with some applicability to government
- Federal Data Strategy and Data Cabinet use private sector practices tailored to government to help assess maturity

According to researchers from the Software Engineering Institute at Carnegie Mellon University, maturity models "provide a way for organizations to approach problems and challenges in a structured way by providing both a benchmark against which to assess capabilities and a roadmap for improving them" (Caralli, Knight, and Montgomery, 2012, p. 2). Maturity models also provide "a set of characteristics, attributes, indicators, or patterns that represent progression and achievement in a particular domain or discipline. The artifacts that make up the model are typically agreed upon by the domain or discipline and are validated through application and iterative recalibration" (Caralli, Knight, and Montgomery, 2012, p. 3). Maturity models use a qualitative or mixed-methods structured approach; for this study, we used them to help describe the level of maturity of various aspects of the DoD's data governance, management, and analytics.

We used four models to help to characterize the DoD's analytic environment for acquisition. Two of the models were created specifically to characterize the maturity of data analytics and management for private-sector companies, while two additional models were tailored to government organizations.

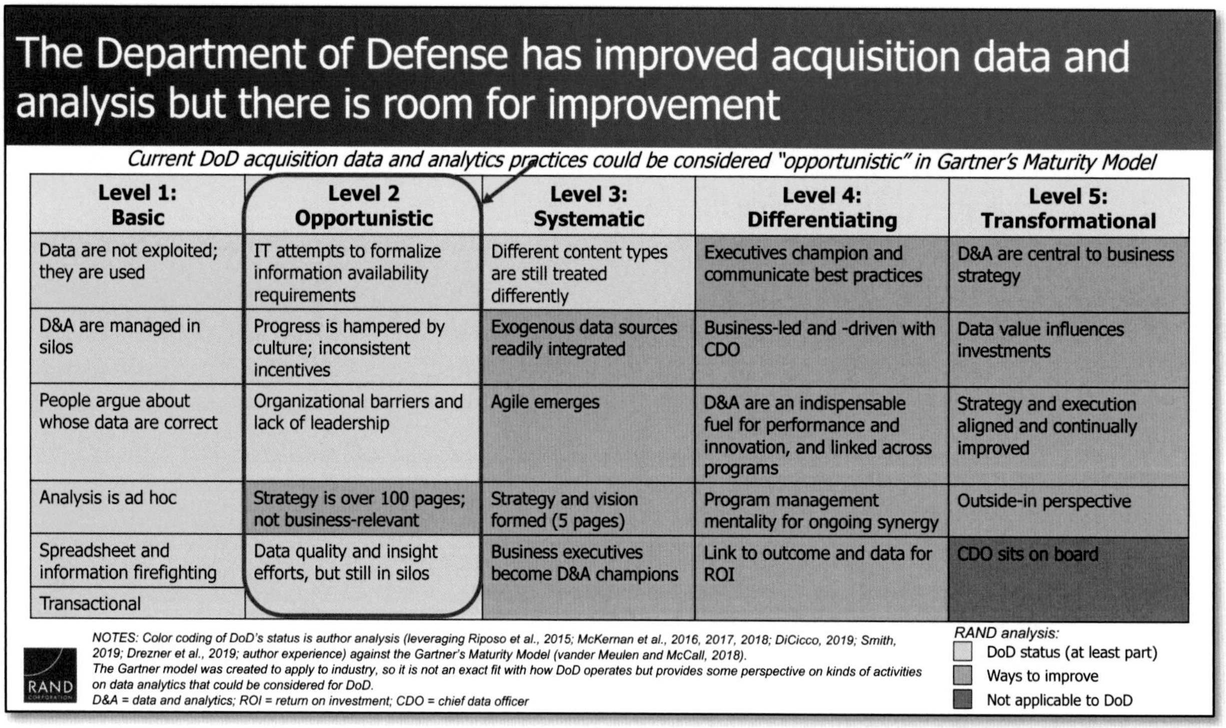

The Department of Defense has improved acquisition data and analysis but there is room for improvement

Current DoD acquisition data and analytics practices could be considered "opportunistic" in Gartner's Maturity Model

Level 1: Basic	Level 2 Opportunistic	Level 3: Systematic	Level 4: Differentiating	Level 5: Transformational
Data are not exploited; they are used	IT attempts to formalize information availability requirements	Different content types are still treated differently	Executives champion and communicate best practices	D&A are central to business strategy
D&A are managed in silos	Progress is hampered by culture; inconsistent incentives	Exogenous data sources readily integrated	Business-led and -driven with CDO	Data value influences investments
People argue about whose data are correct	Organizational barriers and lack of leadership	Agile emerges	D&A are an indispensable fuel for performance and innovation, and linked across programs	Strategy and execution aligned and continually improved
Analysis is ad hoc	Strategy is over 100 pages; not business-relevant	Strategy and vision formed (5 pages)	Program management mentality for ongoing synergy	Outside-in perspective
Spreadsheet and information firefighting	Data quality and insight efforts, but still in silos	Business executives become D&A champions	Link to outcome and data for ROI	CDO sits on board
Transactional				

NOTES: Color coding of DoD's status is author analysis (leveraging Riposo et al., 2015; McKernan et al., 2016, 2017, 2018; DiCicco, 2019; Smith, 2019; Drezner et al., 2019; author experience) against the Gartner's Maturity Model (vander Meulen and McCall, 2018).
The Gartner model was created to apply to industry, so it is not an exact fit with how DoD operates but provides some perspective on kinds of activities on data analytics that could be considered for DoD.
D&A = data and analytics; ROI = return on investment; CDO = chief data officer

RAND analysis:
DoD status (at least part)
Ways to improve
Not applicable to DoD

RAND CORPORATION

The first maturity model that we used to describe the DoD data analytics environment is a model produced by Gartner. This model was used by Gartner for an online survey of private-sector data and analytic practices: "This research was conducted via an online survey in the second quarter of 2017 among Gartner Research Circle members—a Gartner-managed panel composed of IT and business leaders—as well as an external sample source. In total, 196 respondents from [Europe, the Middle East and Africa], [Asia-Pacific,] and North America completed the survey. Respondents spanned 13 vertical industry categories, and revenue categories from 'less than $100 million' to '$10 billion or more'" (vander Meulen and McCall, 2018). According to vander Meulen and McCall at Gartner, "91 percent of organizations [surveyed] have not yet reached a 'transformational' level of maturity in data and analytics, despite this area being a number one investment priority for chief information officers (CIOs) in recent years. . . . The global survey asked respondents to rate their organizations according to Gartner's five levels of maturity for data and analytics. . . . It found that 60 percent of respondents worldwide rated themselves in the lowest three levels" (vander Meulen and McCall, 2018).

Some key findings of the survey are worth noting here:

- Five percent of respondents assessed themselves at level 1, 21 percent at level 2, 34 percent at level 3, 31 percent at level 4, and 9 percent at level 5.
- The top three problems organizations were trying to address were improving process efficiency (54 percent), enhancing customer experience (31 percent), and developing new products (31 percent).
- "Despite a lot of attention around advanced forms of analytics, 64 percent of organizations still consider enterprise reporting and dashboards their most business-critical applications for data and analytics. In the same manner, traditional data sources such as transactional data and logs also continue to dominate, although 46 percent of organizations now report using external data" (vander Meulen and McCall, 2018).
- The three most common barriers to using data analytics are defining data and analytic strategy, determining how to get value from projects, and solving risk and governance issues. The common barriers are what Gartner tends to see in organization at level 2 or 3 maturity (level 5 being the highest) (vander Meulen and McCall, 2018).

This model was built to assess private-sector companies throughout the world, but it offers general qualities that can be applied to the defense acquisition data environment. We used existing research (when available) and our expert judgment and experience to assess the DoD's status against Gartner's maturity model. The DoD acquisition community likely resides at approximately level 2 for most of the characteristics of its data analytics. We base this evaluation on the research that we have done since 2014 for the DoD on acquisition information management and access (see Riposo et al., 2015; McKernan et al., 2016; McKernan et al., 2017; McKernan et al., 2018). Within the DoD acquisition community, the level of maturity also varies across the DoD.

For instance, at level 1, data analytics are managed in silos. In some offices, data are still managed in silos. A good example is the information generated by DCMA, which resides largely within its network and is not easily accessible from outside DCMA. However, a lot of information is stored in centrally located information systems and is being accessed by thousands of DoD users for a variety of purposes. An example of this situation is DAMIR, which contains program data from multiple sources, including three military departments. Despite the progress made to move from silos to centralized information, culture still inhibits sharing, which aligns with level 2. We also know that several organizations are working toward a data strategy, but no DoD-wide acquisition data strategy exists, which also aligns with level 2.

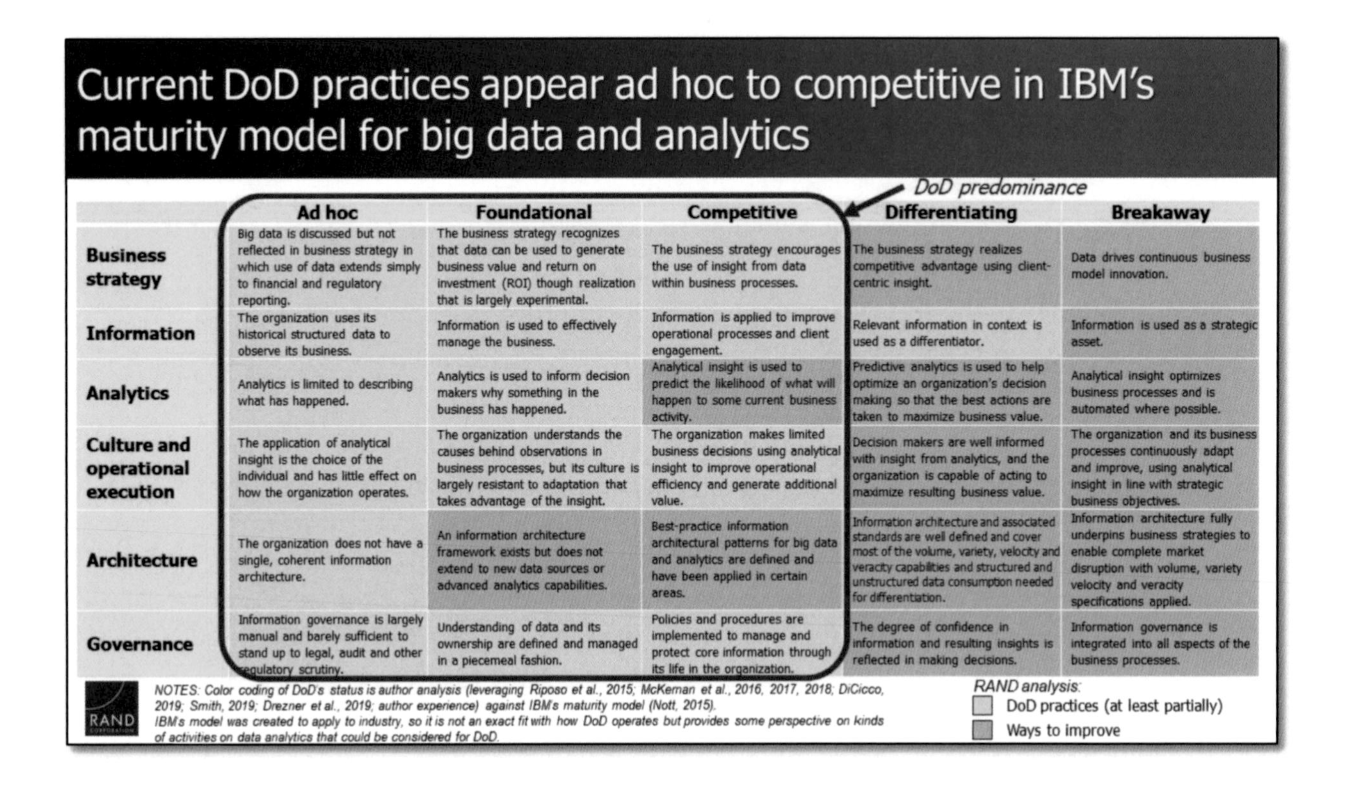

	Ad hoc	Foundational	Competitive	Differentiating	Breakaway
Business strategy	Big data is discussed but not reflected in business strategy in which use of data extends simply to financial and regulatory reporting.	The business strategy recognizes that data can be used to generate business value and return on investment (ROI) though realization that is largely experimental.	The business strategy encourages the use of insight from data within business processes.	The business strategy realizes competitive advantage using client-centric insight.	Data drives continuous business model innovation.
Information	The organization uses its historical structured data to observe its business.	Information is used to effectively manage the business.	Information is applied to improve operational processes and client engagement.	Relevant information in context is used as a differentiator.	Information is used as a strategic asset.
Analytics	Analytics is limited to describing what has happened.	Analytics is used to inform decision makers why something in the business has happened.	Analytical insight is used to predict the likelihood of what will happen to some current business activity.	Predictive analytics is used to help optimize an organization's decision making so that the best actions are taken to maximize business value.	Analytical insight optimizes business processes and is automated where possible.
Culture and operational execution	The application of analytical insight is the choice of the individual and has little effect on how the organization operates.	The organization understands the causes behind observations in business processes, but its culture is largely resistant to adaptation that takes advantage of the insight.	The organization makes limited business decisions using analytical insight to improve operational efficiency and generate additional value.	Decision makers are well informed with insight from analytics, and the organization is capable of acting to maximize resulting business value.	The organization and its business processes continuously adapt and improve, using analytical insight in line with strategic business objectives.
Architecture	The organization does not have a single, coherent information architecture.	An information architecture framework exists but does not extend to new data sources or advanced analytics capabilities.	Best-practice information architectural patterns for big data and analytics are defined and have been applied in certain areas.	Information architecture and associated standards are well defined and cover most of the volume, variety, velocity and veracity capabilities and structured and unstructured data consumption needed for differentiation.	Information architecture fully underpins business strategies to enable complete market disruption with volume, variety velocity and veracity specifications applied.
Governance	Information governance is largely manual and barely sufficient to stand up to legal, audit and other regulatory scrutiny.	Understanding of data and its ownership are defined and managed in a piecemeal fashion.	Policies and procedures are implemented to manage and protect core information through its life in the organization.	The degree of confidence in information and resulting insights is reflected in making decisions.	Information governance is integrated into all aspects of the business processes.

NOTES: Color coding of DoD's status is author analysis (leveraging Riposo et al., 2015; McKernan et al., 2016, 2017, 2018; DiCicco, 2019; Smith, 2019; Drezner et al., 2019; author experience) against IBM's maturity model (Nott, 2015). IBM's model was created to apply to industry, so it is not an exact fit with how DoD operates but provides some perspective on kinds of activities on data analytics that could be considered for DoD.

We also used our prior research, expert judgment, and experience to assess the DoD's status against IBM's maturity model for big data analytics (see Riposo et al., 2015; McKernan et al., 2016; McKernan et al., 2017; McKernan et al., 2018).

IBM's model and its use in a survey are described by Nott:

> According to a recent IBM Institute of Business Value (IBV) study, 63 percent of organizations in 2014 realized a positive return on their analytics investments within a year. That study also noted that 74 percent of respondents anticipate that the speed at which executives expect new data-driven insights will continue to accelerate. . . . The big data analytics maturity model considers not only the technology to lay out a path to success, but more importantly it also takes into account the business factors. (Nott, 2015)

This maturity model has six categories—business strategy, information, analytics, culture and operational execution, architecture, and governance. Note that the Gartner model did not divide characteristics into categories but rather just listed various attributes for each level. Like the Gartner model, the IBM model has five levels of maturity.

The qualities that we have seen in data management in the DoD make it difficult to place the DoD at a specific level of maturity. The DoD has accomplished all features of the ad hoc level but only some of the foundational- and competitive-level attributes. For instance, as mentioned earlier, DoD uses analytics to describe what has happened and why it has happened, as in the case of statutorily mandated Nunn-McCurdy root-cause analyses, but typically does not predict what may happen. Nott notes:

> Mature use of analytics optimizes the business. Organizations will already be reporting to show their financial performance and to demonstrate regulatory compliance, but analytics is necessary to understand why something has happened or to predict what is likely to happen. The resulting insight helps improve customer engagement and operational efficiency. Analytics is used to make data-driven decision making pervasive in an organization, and it requires timely insight in context. (Nott, 2015)

Nott also describes how mature organizations view information and govern that information:

> Information: Use of data to manage the business is the base capability. However, highly mature organizations recognize that data is a first-class, strategic business asset. It comes not only from existing transactional systems—the systems of record—but also from systems that support individual—the systems of engagement—and external data sources. Furthermore, mature organizations provide governed access to data wherever it resides in the organization and are able to give it meaning and context. (Nott, 2015)

The DoD considers information important, but it is not always accessible by those who need it, which may be due to security or cultural reasons, according to Nott:

> Information governance is a critical success factor for big data projects. Policies need to be established and enforced to a degree of confidence in information and so that resulting insights are understood and reflected in decisionmaking efforts. Policies also need to span provenance, currency, data quality, master data and metadata, lifecycle management, security, privacy and ethical use. (Nott, 2015)

In specific instances, the DoD has used its historical data to observe its business. This situation is often seen in analyses conducted by FFRDCs for the DoD. For example, this category includes analyses of cost growth in major weapon systems over time (see Bolten et al., 2008).

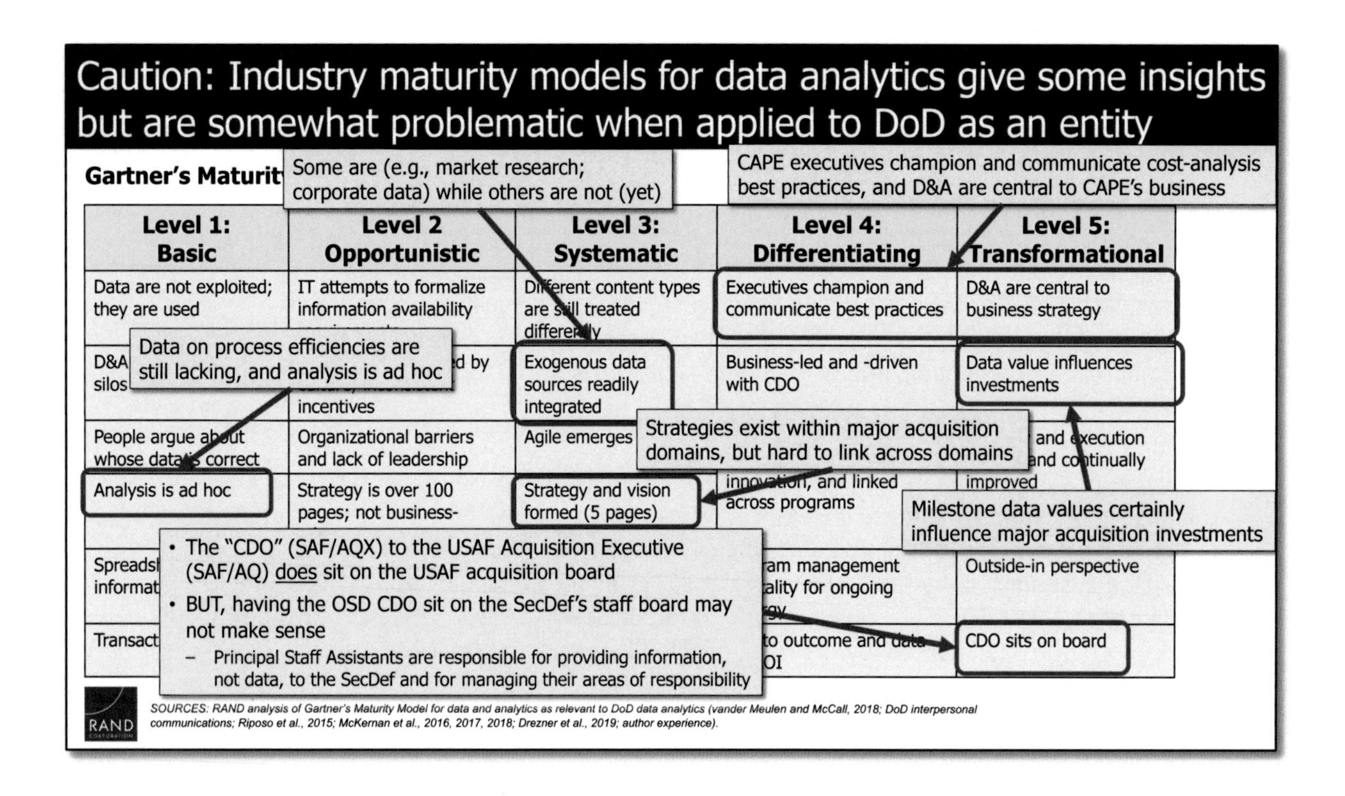

The maturity models for data analytics by Gartner and IBM provide Congress with a better understanding of levels of maturity in this domain (vander Meulen and McCall, 2018; Nott, 2015), as well as where DoD acquisition generally resides, based on our subject-matter expertise in this area. However, we caution that the practices in these models do not directly apply to the DoD as a single entity (with all its different elements and missions) in all instances. For example, a CDO sitting on the Secretary of Defense's staff board might not make sense. Principal staff assistants are responsible for providing information, not data, to the secretary and for managing their areas of responsibility. Another example is CAPE, within OSD. CAPE is performing at a level 4 or 5 (differentiating or transformational), because CAPE executives champion and communicate cost analysis best practices, and data and analytics are central to CAPE's business.

This is not to say that these models might not more directly apply to individual elements of the DoD that have more-similar missions and objectives (e.g., within one of the military departments or within a function, such as personnel and readiness or acquisition and sustainment).

The Federal Data Strategy Development Team, Cross-disciplinary data experts from across the Federal Government, have tailored a model for government use

Objective:	Best Government Practices:									
Govern and Manage Data as a Strategic Asset	Establish Data Governance Structures	Preserve Federal Data	Inventory Data Assets	Publish Data Documentation	Assess Data Maturity	Manage with a Long View	Identify High-Value and Authoritative Data Assets	Align Resources to Value and Authority	Manage High-Value and Authoritative Data Assets	Coordinate Federal Data Assets
Protect and Secure Data	Review Data Releases for Disclosure Risk	Prioritize Data Security	Define Responsibilities for Protecting Confidentiality	Evolve Data Security	Diversify Data Access Methods	Innovate to Enable Safe Use	Preserve Data Integrity	Align Contracts with Data Management Requirements		
Promote Efficient Use of Data Assets	Leverage Data Standards	Explicitly Communicate Allowable Use	Connect Data Functions Across Agencies	Increase Staff Capacity for Data Analysis	Promote Wide Access	Maximize Economic Value through Open Access	Enable Use through Data Platforms	Prevent Monopolization of Federal Data		
	Improve Secure Data Linkage	Prepare to Share	Share Data Across Agencies	Leverage Buying Power	Plan Ahead with Informed Consent	Share Data Between State and Local Governments and Federal Agencies	Recover Allowable Costs			
Build a Culture that Values Data as an Asset	Incorporate Data into Decisionmaking	Promote a Culture That Values Data as an Asset	Connect Federal Spending to Outcomes	Focus on End Uses of Data	Communicate Insights from Data	Plan for Evidence-Building	Innovate with Partners			
Honor Stakeholder Input and Leverage Partnership	Honor Proprietary Interests	Assess the Needs of Stakeholders	Leverage Partnerships	Balance Stakeholder Needs	Monitor and Address Public Perceptions	Allow Amendment	Engage Federal Experts			

Key: Usually; Most of the time; Sometimes; Limited; Never or almost no; n/a to DoD

NOTE: Color coding of DoD's status is author analysis (leveraging Riposo et al., 2015; McKernan et al., 2016, 2017, 2018; DiCicco, 2019; Smith, 2019; Drezner et al., 2019; author experience) against the Federal Data Strategy's draft practices (Office of Management and Budget, undated).

RAND

DoD follows some best practices, but consistency improvements are ongoing

A cross-disciplinary set of data experts from across the federal government tasked with creating a coordinated approach to federal data use and management, the Federal Data Strategy Development Team, has tailored a maturity model for government use. The model is based on data strategy practices identified by the Federal Data Strategy Development Team:

> [The] Federal Data Strategy will consist of principles, practices, and action steps to deliver a consistent and strategic approach to federal data stewardship, access, and use. . . . The practices are designed to inform agency actions on a regular basis, to be continually relevant, and to be sufficiently general so as to broadly apply at all federal agencies and across all missions. The practices represent aspirational goals that, when fully realized, will continually challenge and guide agencies, practitioners, and policymakers to improve the government's approach to data stewardship and the leveraging of data to create value. . . . In addition to applying across government, the strategy and its practices apply across the data lifecycle, which can be depicted in stages: creation or collection or acquisition; processing; access; use; dissemination; and storage and disposition. (U.S. Office of Management and Budget, undated)

The draft practices are grouped according to five broad objectives:

- Govern and manage data as a strategic asset.
- Protect and secure data.
- Promote efficient use of data assets.
- Build a culture that values data as an asset.
- Honor stakeholder input and leverage partners.

As with the Gartner and IBM models, we used our knowledge from prior studies to illustrate the practices in which the DoD is engaged. For instance, it appears that the DoD is engaging in the governing and management of data as a strategic asset more than it is promoting efficient use or building a culture that values data as an asset. The DoD also spends a lot of time protecting and securing data but spends less time honoring stakeholder input and leveraging partnerships. For example, the DoD acquisition community has cataloged some but not all data. Although significant and useful, this effort is not yet complete. It is still very difficult for an analyst covering cross-functional issues to know what data exist and how to access the data, to be able to utilize the data for analysis.

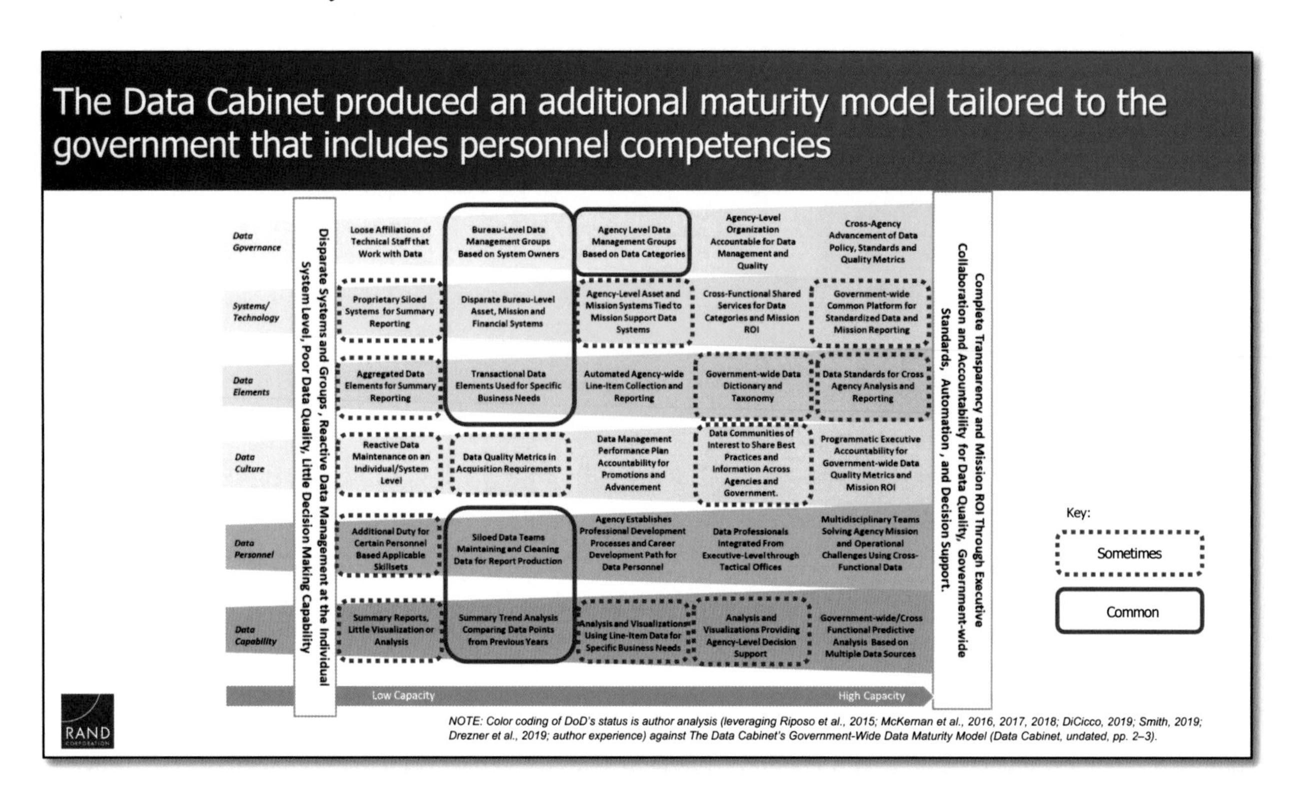

NOTE: Color coding of DoD's status is author analysis (leveraging Riposo et al., 2015; McKernan et al., 2016, 2017, 2018; DiCicco, 2019; Smith, 2019; Drezner et al., 2019; author experience) against The Data Cabinet's Government-Wide Data Maturity Model (Data Cabinet, undated, pp. 2–3).

The Data Cabinet, which is another federal government–level group of experts, constructed an additional maturity model that reflects data-related practices that should be considered in government. The purpose of the model is the following:

- Help agencies with a high-level assessment of current capabilities and supporting processes.
- Help with strategic communication between agency data professionals and agency leadership.
- Provide a common language and framework to help promulgate common solutions and best practices across federal agencies toward advancing data-driven decisionmaking (Data Cabinet, undated).

The primary difference between this model and the others is that it includes a category for data personnel maturity. An important consideration in data management is the execution of sound data practices by the workforce. Sound practices will not be used without a knowledgeable workforce. Along with a challenging data culture, acquiring, training, and retaining data talent might be one of the DoD acquisition community's greatest challenges. The DoD acquisition data community still appears to be using siloed data teams for maintaining and cleaning data, rather than establishing agency-wide professional development and career development paths for data personnel.

In terms of data governance, this capability generally still resides at the local level or within a functional domain (e.g., program information, contracting, contract management, or auditing). Information systems generally are still disparate and at the bureau level, but pockets of higher-level maturity are using application programming interfaces (APIs) and are attempting to create a common data framework for acquisition program data. The practice of creating a list of data elements and definitions is also becoming more common, along with defining and utilizing authoritative sources.

The commercial sector offers some lessons learned from its challenges for DoD's current acquisition analytic environment

- McKinsey & Company recently identified signals that private sector organizations may be unsuccessful in data analytics implementation. These include:
  - The executive team doesn't have a **clear vision** for its advanced-analytics programs
  - No one has determined the **value** that the initial **use cases** can deliver in the first year
  - There's no **analytics strategy** beyond a few use cases
  - **Analytics roles**—present and future—are poorly defined
  - The organization **lacks analytics translators**
  - **Analytics capabilities are isolated from the business**, resulting in an ineffective analytics organization structure
  - Costly **data-cleansing** efforts are started *en masse*
  - **Analytics platforms** are not built to purpose
  - Nobody knows the **quantitative impact that analytics is providing**
  - No one is hyperfocused on **identifying potential ethical, social, and regulatory implications** of analytics initiatives
- Despite having a different overall objective in DoD as compared to increasing profit in the commercial sector, these "red flags" are applicable to DoD

Source: McKinsey & Company (Fleming et al., 2018).

The private sector offers some lessons learned from its challenges for the DoD's current acquisition analytic environment. For example, McKinsey & Company has spent some time trying to identify success and failure in companies' investments in analytics in artificial intelligence (AI):

> [T]here's one upside to the growing list of misfires and shortfalls in companies' big bets on analytics and AI. Collectively, they begin to reveal the failure patterns across

organizations of all types, industries, and sizes. We've detected what we consider to be the ten red flags that signal an analytics program is in danger of failure. In our experience, business leaders who act on these alerts will dramatically improve their companies' chances of success in as little as two or three years. (Fleming et al., 2018)

The following is a list of the red flags:

- The executive team does not have a clear vision for its advanced-analytic programs.
- No one has determined the value that the initial use cases can deliver in the first year.
- There is no analytic strategy beyond a few use cases.
- Analytic roles—present and future—are poorly defined.
- The organization lacks analytic translators.
- Analytic capabilities are isolated from the business, resulting in an ineffective analytic organization structure.
- Costly data-cleansing efforts are started en masse.
- Analytic platforms are not built to purpose.
- Nobody knows the quantitative impact that analytics are providing.
- No one is hyperfocused on identifying potential ethical, social, and regulatory implications of analytic initiatives (Fleming et al., 2018).

These lessons learned are applicable to the DoD. In particular, the importance of understanding the value of analytics to the DoD before investing resources is critical, as is a clear vision from the DoD leadership.

Business-sector data science best practices may offer solutions but should be carefully reviewed for applicability to DoD's environment

- Organizations should have an overall "data vision" and analytic strategy
- Organizations should aim to standardize data collection and procedures at an enterprise level
  - Provide awareness of what data is being collected
  - Avoid duplicative data collection efforts resulting from "siloed" data
  - Allow the right people access to the data they need
- Ensuring good data quality is key to good analysis
- Data storage should be designed to meet the organization's need
  - Needs depend on the volume of data, data structure(s), and access requirements
  - Data lakes (and similar) are a current data storage best practice, but DoD would need to consider the appropriateness of data lakes as a storage strategy
- Responsibility for data and analytics resides within the business units, with coordination and support from headquarter (centralized) organizations
  - Better enables collaboration between subject matter experts, data scientists/analysts, and decision makers

Private-sector data best practices for analytics may offer solutions but should be carefully reviewed for applicability to the DoD environment. One such practice is that organizations should have an overall data vision and analytic strategy:

> Many businesses, seeing digital opportunities (and digital competition) in their sectors, rush to invest without a considered, holistic data strategy. They either focus on the technologies alone or address immediate, distinct use cases without considering the mid- to long-term creation of sustainable capabilities. This goes some way toward explaining why a 2017 McKinsey Global Survey found that only half of responding executives report even moderate effectiveness at meeting their analytics objectives. (McKinsey Analytics, 2018, p. 59)

The DoD can learn from the challenges that private-sector firms have already experienced in regard to establishing a strategy before employing widespread data analytics.

Organizations should aim to standardize data collection and procedures at an enterprise level. This effort would help improve awareness of what data are being collected, avoid duplicative data collection efforts resulting from siloed data, and allow the right people access to the data they need. Ensuring good data quality is also key to good analysis. For example,

> IBM estimates that a staggering 70% of the time spent on analytic projects is concerned with identifying, cleansing, and integrating data due to the difficulties of locating data that is scattered among many business applications, the need to reengineer and reformat it in order to make it easier to consume, and the need to regularly refresh it to keep it up-to-date. (Terrizzano et al., 2015, p. 1)

Good data governance and management practices would help mitigate some of the time spent upfront.

Data storage should be designed to meet the organization's needs. An organization's needs depend on the volume of data, data structures, and access requirements. Data lakes (and similar) are a common data storage practice, but the DoD would need to consider the appropriateness of data lakes as a storage strategy.[8]

Responsibility for data and analytics resides within the business units, with coordination and support from headquarter (centralized) organizations. This practice better enables collaboration among subject-matter experts, data scientists and analysts, and decisionmakers.

---

[8] According to Amazon Web Services (undated): "A data lake is a centralized repository that allows you to store all your structured and unstructured data at any scale. You can store your data as-is, without having to first structure the data, and run different types of analytics—from dashboards and visualizations to big data processing, real-time analytics, and machine learning to guide better decisions."

Spending on the Analytic Workforce

What's in this Briefing

- Background
- Approach and methodology
- Findings on congressional questions
  - Q1–Q2: Extent of, and potential to increase, DoD data analytics capabilities
    - *Map decisions to analyses and data; identify gaps*
    - *Data analytics availability and use on selected programs*
    - *DoD's information systems that enable data analytics*
    - *DoD's status against maturity models*
    - *Spending on the analytic workforce*
    - *Spending on IT data and analysis systems*
    - *Example limits in data availability and analytic challenges from this study*
  - Q3: R&D funding for data analytics capabilities
  - Q4: Private-sector best practices for efficient data collection and delivery
  - Q5: Anonymized data sharing with researchers and analysts
  - Q6: Appropriate data analytics and methods courses at defense training institutions
- Conclusions

RAND

Next, we will help quantify the extent of the DoD's analytic capabilities by estimating the annual spending on analytic elements of the AWF.

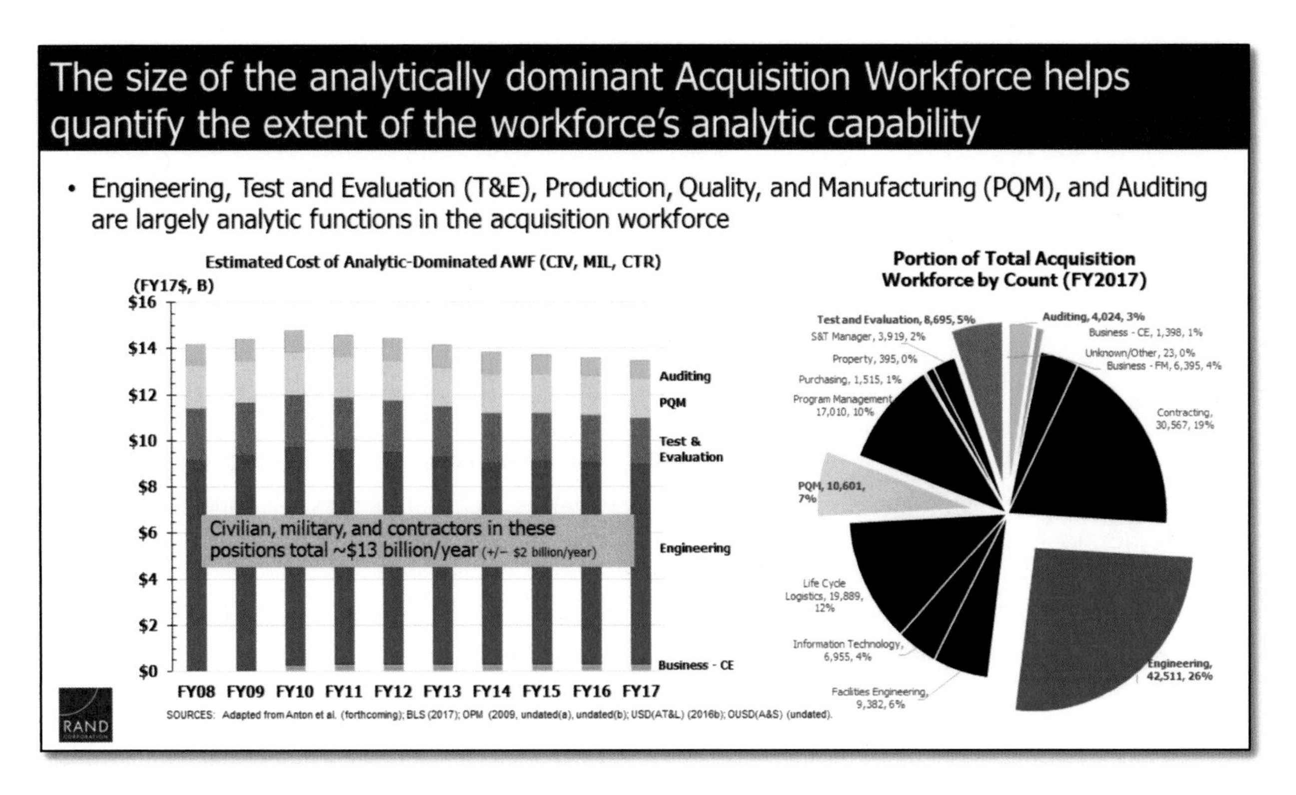

NOTE: CIV = civil servant; MIL = military personnel; CTR = contractor.

We adapted the results from Anton et al. (forthcoming) to extract the portion of the AWF that is predominantly data analytic. OSD's Human Capital Initiative provides official tracking and data on the AWF (Office of the Under Secretary of Defense for Acquisition and Sustainment, undated). Human Capital Initiative breaks down the workforce into 14 AWF career fields for civil servants and military personnel. These categories are engineering, contracting, life-cycle logistics, PM, PQM, facilities engineering, T&E, IT, business financial management, auditing, S&T manager, purchasing, business cost estimating, and property.

The business cost estimating, engineering, T&E, PQM, and auditing functions are predominantly data analytics in nature, as can be seen by some of the formal definitions of these functions:

- *Cost analysis* is an "analysis and evaluation of each element of cost in a contractor's proposal to determine reasonableness" (DAU, 2017a).
- *Business, cost estimating, and financial management* consists of "[m]anagement of acquisition funds including, but not limited to: cost estimating; formulation of input for the Program Objectives Memorandum (POM), the budget, and other programmatic or financial documentation of the Planning, Programming, Budgeting, and Execution (PPBE) process; and budget execution (paying bills)" (DAU, 2017a).
- *Cost estimate* is an "estimate of the cost of an object, commodity, weapon system, or service resulting from an estimating procedure or algorithm. A cost estimate has "context," that is, whether it is the cost to develop and/or procure, and/or to support and/or maintain the item of service and whether it is an incremental, total or Life Cycle

Cost, or some other cost perspective. A cost estimate may constitute a single value or a range of values" (DAU, 2017a).

- *Systems engineering* is an "interdisciplinary approach and process encompassing the entire technical effort to evolve, verify and sustain an integrated and total life cycle balanced" (DAU, 2017a). Also, engineering is heavily quantitative, analyzing a wide range of data, including specifications, designs, system performance, production, and planning of hardware and software systems.
- *Developmental test and evaluation* is "(1) Any testing used to assist in the development and maturation of products, product elements, or manufacturing or support processes. (2) Any engineering-type test used to verify status of technical progress, verify that design risks are minimized, substantiate achievement of contract technical performance, and certify readiness for initial operational testing. Developmental tests generally require instrumentation and measurements and are accomplished by engineers, technicians, or soldier operator-maintainer test personnel in a controlled environment to facilitate failure analysis" (DAU, 2017a).
- *Operational test and evaluation* consists of a "field test, under realistic conditions, of any item (or key component) of weapons, equipment, or munitions for the purpose of determining the effectiveness and suitability of the weapons, equipment, or munitions for use in combat by typical military users; and the evaluation of the results of such tests" (DAU, 2017a).
- *Quality control* is the "system or procedure used to check product quality throughout the acquisition process" (DAU, 2017a).
- *Quality audit* is a "systematic examination of the acts and decisions with respect to quality in order to independently verify or evaluate the operational requirements of the quality program or the specification or contract requirements for a product or service" (DAU, 2017a). These checks and examinations analyze data on system performance and production operations.
- *Audit* is a "[s]ystematic examination of records and documents to determine adequacy and effectiveness of budgeting, accounting, financial, and related policies and procedures; compliance with applicable statutes, regulations, policies, and prescribed procedures; reliability, accuracy, and completeness of financial and administrative records and reports; and the extent to which funds and other resources are properly protected and effectively used" (DAU, 2017a).

As a result, measuring the size of those five Human Capital Initiative categories provides a rough estimate of the analytic capability in terms of workforce. Anton et al. (forthcoming) estimated the monetary cost of the AWF numbers using cost and demographic data from OPM (2009, undated-a, undated-b), the Defense Manpower Data Center, and the U.S. Bureau of Labor Statistics (undated). The previous slide shows that these five elements of the AWF constitute about $13 billion per year (plus or minus $2 billion) of data analytics capability for the DoD in FY 2017, a slight downward trend in real (adjusted) dollars.

Of course, not everything these workers do is analytic, and aspects of the functions in the other AWF categories are data analytic, but this assessment provides a first-order parametric approximation of the analytic workforce capability in the DoD.

Broader perspectives help to expand the understanding of workforce capabilities

Other workforce-related aspects besides raw size reflect analytic extent:

- Training and educational opportunities provided to the workforce
- Dedicated analysis centers
  - Examples: OSD CAPE, PDASD (Acquisition Enablers); analysis shops supporting the Service Acquisition Executives
- Establishment of CDOs in the OSD (2018), Air Force (2017), and Army (2009)

NOTE: PDASD = principal deputy assistant secretary of defense.

In addition to this macro measure of certain analytic elements in the AWF, it is worth noting three other workforce-related elements. First, the extent of data analytics training for the AWF provides insight into the skills and expertise across the AWF. This factor is analyzed later in the report, in response to Congress's sixth question.

Second, we note the continued existence and importance of having data analytics organizations to support acquisition leadership. Examples for programmatic information are OSD CAPE, the analytic divisions under PDASD for acquisition enablers in the newly organized OUSD(A&S), and the analytic divisions supporting the SAEs. Other positions abound in specific acquisition functions (e.g., T&E laboratories and centers), but it is beyond the scope of this effort to try to build a comprehensive list of such organizations.

Finally, we also note the establishment of CDOs in OSD (2018), the Air Force (2017), and the Army (2009). The Navy is still weighing the utility of establishing a CDO separate from the CIO. Although these CDOs have broader responsibilities that extend beyond acquisition, their existence reflects the DoD's recognition that data need to be managed strategically if they are to enable analysis and inform decisions. The DoD's CDOs are relatively new and still need to consider appropriate measures for the DoD's large collection of information (including acquisition data). Industry integrates corporate data and control under a CDO, but this practice needs to be applied carefully to the DoD, given the stovepiped environment and uneven data practices. Industry best practice also locates responsibility for data analytics in the business unit that is using the data analysis, rather than a separate headquarter organization, such as a CDO or CIO, so this practice would need to be considered as the CDO role matures within the DoD.

Spending on Major IT Data and Analysis Systems

What's in this Briefing

- **Background**
- **Approach and methodology**
- **Findings on congressional questions**
  - Q1–Q2: Extent of, and potential to increase, DoD data analytics capabilities
    - *Map decisions to analyses and data; identify gaps*
    - *Data analytics availability and use on selected programs*
    - *DoD's information systems that enable data analytics*
    - *DoD's status against maturity models*
    - *Spending on the analytic workforce*
    - *Spending on IT data and analysis systems*
    - *Example limits in data availability and analytic challenges from this study*
  - Q3: R&D funding for data analytics capabilities
  - Q4: Private-sector best practices for efficient data collection and delivery
  - Q5: Anonymized data sharing with researchers and analysts
  - Q6: Appropriate data analytics and methods courses at defense training institutions
- **Conclusions**

Another way to document the current extent of DoD data analytics capabilities is to measure how much is being spent on IT systems that support acquisition processes, including enabling data collection and analysis.

Another way to document the status quo is to estimate how much is being spent on analytic capabilities

- A rough estimate can be constructed through using spending on IT information systems and justifications in budgets:
  - Summarized annual IT investments in *Acquisition* and in *Logistics and Supply-Chain Management* as reported in SNaP-IT
- ~$3 billion is spent annually on IT systems for processing, data collection, and analysis
  - ~$0.5 billion/year for *acquisition*
  - ~$2.5 billion/year for *logistics and supply-chain management*

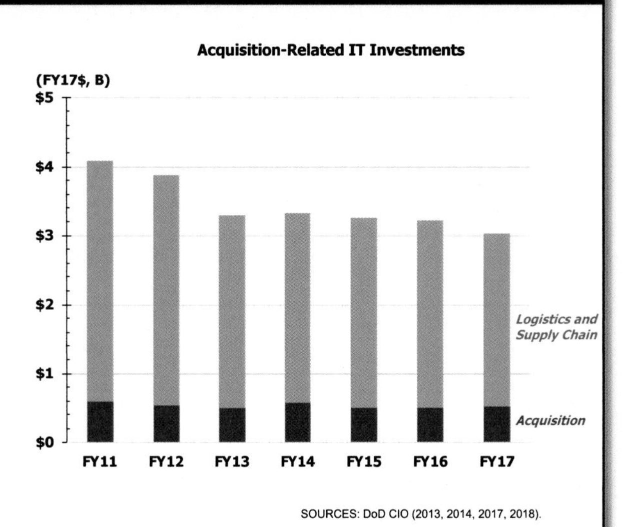

A rough estimate of spending was constructed by summarizing annual IT investments reported yearly in the PB request under the SNaP-IT exhibit. SNaP-IT is a database application and report that is used to plan, coordinate, edit, publish, and disseminate IT budget justifications for internal use and for Congress (see USD[Comptroller], 2017). Within the SNaP-IT database are metadata that identify the primary function that each IT system supports. Given our broad definition of *acquisition*, the two relevant functions in SNaP-IT are *acquisition* and *logistics and supply-chain management*. These major IT systems support specific processes and have data collection capabilities. Examples include contract writing and tracking systems, which record data on individual contracts and contracting actions to acquire goods and services. Some have analytic layers to enable data analytics directly on the data within the system. We cannot tell from the SNaP-IT data the extent of functionality associated with process support, data collection, and analytics, but these systems reflect best practices of collecting acquisition data directly from acquisition operations and functions rather than requiring manual resubmission of selected data into a separate system. Measuring SNaP-IT systems also give rough insight into the extent of the DoD's capabilities in collecting data from acquisition operations.

The previous slide summarizes spending and trends on major IT systems from FY 2011 to FY 2017 (DoD CIO, 2013, 2014, 2017, 2018). It shows that the DoD currently spends about $3 billion on IT investments in the categories of acquisition and logistics and supply-chain management systems. About half a billion are for acquisition systems, and about $2.5 billion are for logistics and supply-chain management systems. These expenditures have decreased in terms of real (adjusted) FY 2017 dollars from FY 2011, when the total spending was about $4 billion.

The *acquisition* subsegment of IT systems cuts across acquisition functions

Top 60 % (by dollars) of *Acquisition* Subsegments for FY 2017

	FY 2017, thousands	% of total	Cum %	Investment Name	Function
1	40,690	7.9%	8%	SPAWAR Acquisition Capabilities	PM
2	30,863	6.0%	14%	Science & Technology	RDT&E
3	28,862	5.6%	20%	Integrated Award Environment	Personnel
4	19,868	3.9%	23%	Army Contract Writing System	Contracting
5	14,235	2.8%	26%	Standard Procurement System	Contracting
6	14,014	2.7%	29%	SPAWAR Systems Acquisition Management Capabilities	PM
7	13,536	2.6%	32%	Wide Area Workflow	Contracting
8	11,622	2.3%	34%	Federal Mall	Procurement
9	11,455	2.2%	36%	NMCI Enterprise Service Tools	Personnel
10	10,745	2.1%	38%	Software Engineering Institute	Software
11	9,484	1.9%	40%	AcqBusiness	Networks
12	8,766	1.7%	42%	Integrated Defense Enterprise Acquisition System	Telecommunications
13	8,281	1.6%	43%	Standard Procurement System	Procurement
14	8,178	1.6%	45%	Electronic Document Access	Contracting
15	7,626	1.5%	46%	Enterprise Business System	Logistics modernization
16	6,973	1.4%	48%	Virtual Contracting Enterprise	Contracting
17	6,953	1.4%	49%	Bureau Of Naval Personnel (BUPERS) Information Technology Support	Financial management
18	6,938	1.4%	51%	Contracting Information Technology	Contracting
19	6,469	1.3%	52%	DISA Storefront	Procurement
20	6,413	1.3%	53%	SEAPORT	IT
21	6,343	1.2%	54%	Software Engineering Institute (SEI) Applied Research	Software
22	6,180	1.2%	56%	Federal Mall	Procurement
23	5,984	1.2%	57%	Project Management Resource Tools	PM
24	5,732	1.1%	58%	Science & Technology	RDT&E
25	5,323	1.0%	59%	Mechanization Of Contract Administration Services	Contracting
26	5,318	1.0%	60%	SPAWAR Systems Acquisition Management Capabilities	PM
total:	512,402		100%		

SOURCE: DoD CIO (2018).

26 out of n=349 systems

RAND

A quick analysis of the top 60 percent (by dollars) of the acquisition subsegment of major IT systems (DoD CIO, 2018) shows that these subsegments cut across a number of acquisition functions, including PM, RDT&E, contracting, procurement, and software and IT acquisitions. Also, the size (in dollars) of these investments is skewed, with a few large systems consuming most of the budget and a large number of decliningly smaller systems trailing. In the acquisition subsegment shown in the slide, these 26 systems (out of the 349 with budget in FY 2017) involve 60 percent of the approximately $0.5 billion budget.

The *logistics and supply chain* subsegment of IT systems cuts across acquisition functions

Top 50 % (by dollars) of *Logistics and Supply Chain* Subsegments for FY 2017

	FY 2017, thousands	% of total	Cum %	Investment Name	Function
1	260,002	10.3%	10%	Enterprise Business System	Materiel and tool tracking
2	131,839	5.2%	16%	Global Combat Support System—Army	OSHA data and trending
3	121,762	4.8%	20%	Logistics Modernization Program Inc.1 Production Baseline	Issue tracking and metrics
4	99,152	3.9%	24%	Global Combat Support System—Army	HR automation and tracking
5	61,174	2.4%	27%	Global Combat Support System—MC/LCM	Network funding reports
6	43,659	1.7%	28%	Navy Enterprise Resource Planning	Logistics and financial mgmt
7	41,325	1.6%	30%	Distribution Standard System	Telecom invoices and billing
8	41,217	1.6%	32%	Commissary Advanced Resale Transaction Sys Replacement/Mod	Telecom mgmt
9	39,092	1.5%	33%	Global Combat Support System—Army	Lab RDT&E networking
10	34,130	1.4%	35%	Global Combat Support System—Army	Config & engr. data management
11	32,767	1.3%	36%	Logistics Information Warehouse	Combat service support management
12	31,640	1.3%	37%	Base Level Support Application	Base telecom management
13	29,284	1.2%	38%	Navy Enterprise Resource Planning	Log. and financial management
14	26,887	1.1%	39%	Federal Logistics Information System	Log. and engineering support
15	26,224	1.0%	40%	Navy Maritime Maintenance Enterprise Solution	IT mgmt
16	24,230	1.0%	41%	Global Combat Support System—Army	Config & engineering data management
17	24,151	1.0%	42%	DoD Data Services	Financial management data sharing
18	22,733	0.9%	43%	Integrated Logistics Systems—Supply	Accounting
19	21,988	0.9%	44%	Enterprise Business Solution	PPBE, manpower, and personnel mgmt
20	21,549	0.9%	45%	Maintenance Figure Of Merit	Financial management
21	19,314	0.8%	46%	Enterprise Business System	Config. management
22	18,895	0.7%	46%	NAVAIR Logistics Capabilities	Debt mgmt
23	18,836	0.7%	47%	Defense Personal Property System	MilPer financial and leave management
24	18,615	0.7%	48%	Logistics Modernization Program Increment 1 Production Baseline	Data sharing
25	18,538	0.7%	49%	RF In-Transit Visibility	Logistics and financial management
26	17,458	0.7%	49%	Commissary Advanced Resale Transaction System	Fin. management data sharing
27	16,514	0.7%	50%	Naval Tactical Command Support System	Procurement reporting and tracking
total:	2,527,187		100%		

SOURCE: DoD CIO (2018).

27 out of n=1,002 systems

RAND

Similarly, a quick analysis of the top 50 percent (by dollars) of the logistics and supply-chain management subsegment of IT systems (DoD CIO, 2018) shows that they cut across a number of functions, including support, management, tracking, or reporting of data for IT, telecommunication systems, logistics, and materiel.[9] Note that some of these systems involve logistics beyond that associated with acquired systems (e.g., the Defense Personal Property System manages movement of personal household goods for military personnel, not movement of parts and supplies). Still, this rough estimate gives some sense of the general magnitude of IT data tracking, management, and reporting related to logistics and supply chains.

As with the acquisition subsegment, the size (in dollars) of these investments is skewed, with a few large systems consuming most of the budget and a large number of decliningly smaller systems trailing. In the logistics and supply-chain management segment shown in the slide above, these 27 systems (out of the 1,002 with budgets in FY 2017) involve 50 percent of the approximately $2.5 billion budget.

[9] We showed only 50 percent instead of 60 percent, as with the prior slide, because we could not fit 60 percent on this slide.

Using the measure of how much is being spent on IT information systems that support acquisition has strengths and limitations

- This measure captures the kind of efficient data collection from acquisition functions that is a best practice of the kind that Congress asked about in their Question 4
- However, there are strengths and limitations to using this measure
  - The split between collection, processing, and creating tools for analysis is not easily separable in these data
    - *In most cases, data collection, archiving, and processing are primary functions/purposes of these IT investments, while analysis is secondary (McKernan et al., 2017)*
  - It is difficult to precisely assess the core motivation of each IT investment without very specific information on each investment

Although providing insight, this methodology has both strengths and weaknesses. These IT systems include a mix of acquisition process support, data collection, data archiving, and analytic layers. These data help measure the underlying enabling capability that ultimately informs acquisition decisions during execution, management, and oversight. Thus, this methodology includes the kind of efficient data collection from acquisition functions that represents the kind of best practice Congress specifically asks about in question 4. However, the split among collection, processing, and creating tools for analysis is not clearly separable in these data. Prior research found that data collection, archiving, and processing are generally primary functions or purposes of these IT investments, while analysis is secondary (McKernan et al., 2017). Also, although these IT systems relate primarily to acquisition, logistics, and supply-chain management, more analysis is needed to identify what specific functions are supported.

Limitations in the Available Data and Analytic Challenges in Answering Congress's Questions

What's in this Briefing

- Background
- Approach and methodology
- Findings on congressional questions
  - Q1–Q2: Extent of, and potential to increase, DoD data analytics capabilities
    - *Map decisions to analyses and data; identify gaps*
    - *Data analytics availability and use on selected programs*
    - *DoD's information systems that enable data analytics*
    - *DoD's status against maturity models*
    - *Spending on the analytic workforce*
    - *Spending on IT data and analysis systems*
    - *Example limits in data availability and analytic challenges from this study*
  - Q3: R&D funding for data analytics capabilities
  - Q4: Private-sector best practices for efficient data collection and delivery
  - Q5: Anonymized data sharing with researchers and analysts
  - Q6: Appropriate data analytics and methods courses at defense training institutions
- Conclusions

The difficulty of this study (limits in data available and analytic challenges) illustrates the DoD's larger data analytics challenges

Congress' questions:

1. Extent
2. Potential
3. R&D funding
4. Data collection and delivery
5. Data exposure
6. Training

- Some data are simply not tracked
  - Activity-based detailed personnel tasks
    - *Labor is tracked by OPM career field, not what tasks people do. This limits our understanding of tasks and any waste or efficiencies.*
  - Detailed investments
    - *Budget details are hard to access and only provide summaries. Task details are unavailable outside supervisory levels.*
- Most data are not available centrally
- Analysis opportunities are hard to extrapolate from Industry examples
  - Government has different objectives than companies
  - Advanced analytics often depend on large training sets which only exist in some parts of DoD acquisition
- Anonymization or blind-use techniques and policies have not been developed for sharing sensitive data

The limitations on our study (in terms of both data and analytic methods to answer Congress's questions) help illustrate the DoD's larger data analytics challenges. For example, some data are simply not tracked. Labor is tracked by the OPM career field, not by what tasks people do. This practice limits our ability to understand tasks and any waste or efficiencies (i.e., it is difficult, if not impossible, to retrieve activity-based detailed personnel tasks). In addition, details about the DoD's analytic investments are limited. Budget details are hard to access and provide only summaries, and task details are unavailable outside supervisory levels. The data required to answer Congress's questions are not available centrally, and a data request would likely result in obtaining data that are not comparable across offices.

Although we have tried to integrate private-sector lessons learned into this analysis, only some of the examples are transferrable to DoD acquisition. The government's objectives differ from those of companies. Advanced analytics often depend on large training sets that exist only in some parts of DoD acquisition. We, therefore, cannot easily suggest that the DoD specifically follow industry best practices in all cases (more research would be required to assess promising applications, perhaps based on theoretical features of problem sets and analytic tools).

Finally, anonymization or blind-use techniques and policies have not been developed for sharing sensitive data, nor may they be appropriate, given the need to understand the metadata for the analyses.

Question 3. Research and Development Funding for Analytic Capabilities for Acquisition

Program-Element Research, Development, Testing and Evaluation Budgets

What's in this Briefing

- Background
- Approach and methodology
- Findings on congressional questions
  - Q1–Q2: Extent of, and potential to increase, DoD data analytics capabilities
  - Q3: R&D funding for data analytics capabilities
    - *Program-element RDT&E budgets*
    - *IT system RDT&E budgets*
    - *SBIR and STTR solicitations*
  - Q4: Private-sector best practices for efficient data collection and delivery
  - Q5: Anonymized data sharing with researchers and analysts
  - Q6: Appropriate data analytics and methods courses at defense training institutions
- Conclusions

Congress's third question is about "the amount of funding for intramural and extramural research and development activities to develop and implement data analytics capabilities in support of improved acquisition outcomes" (U.S. House of Representatives, 2016, p. 1125). We used budgetary documentation to provide a macro view of what is being spent on analytic RDT&E (at the level of fidelity of DoD budgets) for PEs and for IT systems. We also reviewed the DoD's recent SBIR and STTR solicitations to augment these assessments.

An approach to estimate R&D funding for acquisition analytics is to mine the President's Budget Request for specific investments

1. Searched DoD PB 2019 (Exhibit R-2, RDT&E Budget Item Justification) for analytic keywords supporting acquisition decision making for first filter
   - Analysis, analytic, artificial intelligence, big data, business intelligence, data mining, data science, data scientist, decision support, machine learning, predictive
2. Conducted lexical content analysis to search for potentially relevant phrases
3. Read and assessed relevant Program Elements (PEs) by estimated fraction of dollars associated with data analytics:
   - *All* or *Nearly All*
   - *Moderate* [somewhere Between *Nearly All* and a *Small Part*, exclusive (i.e., there's no overlap with "All or Nearly All and Small Part")]
   - *Small Part*

As part of the research effort to describe the status quo, the research team examined the FY 2019 PB for RDT&E, with an eye toward uncovering specific data analytics efforts supporting acquisition decisionmaking. The annual DoD budget process requires formulation and submission of budget requests with formatted presentation and justification of those requests to Congress (USD[Comptroller], 2008, p. 1-7). The primary data element of the budget requests is the PE, and it is within each PE that RDT&E efforts are organized, budgeted, and reviewed (USD[Comptroller], 2017, p. 5-6). PE budget data are compiled into annual justification books (USD[Comptroller], 2017, p. 5-7) with the following exhibits:

- R-2 RDT&E budget item justification
- R-2a RDT&E project justification
- R-3 RDT&E project cost analysis
- R-4 RDT&E program schedule profile
- R-4a RDT&E program schedule detail (USD[Comptroller], 2017, p. 5-7).

The accompanying descriptive language within the budget item justifications (R-2s) provides summary descriptions of varying levels of detail. When indications of data analytics efforts were found within this language, an associated funding level was also provided.

Our approach for estimating R&D funding for acquisition analytics is to mine the PB request for specific investments. Budget justification books are publicly available for review at links from the office of the USD(Comptroller) website (undated). Additionally, the Defense Technical Information Center provides an unrestricted DoD investment budget search (see Defense Technical Information Center, undated). Additional features (e.g., the ability of filtering search requests by PB year or component) provide the ability to examine individual PE justification exhibits apart from the combined component justification book. Keyword searching was chosen as a more efficient alternative to reading all the exhibits.

The research team first searched the exhibits for a condensed list of associated keywords supporting analytic decisionmaking:

- analysis
- analytic
- artificial intelligence
- big data
- business intelligence
- data mining
- data science
- data scientist
- decision support
- machine learning
- predictive.

As expected, the keywords show a frequency disparity. The word *analysis* is found in 714 individual PE R-2s, while 16 R-2s contain *business intelligence*. The research team used text analysis software (RAND-Lex) to expand the search methodology. Text analytic tools support a variety of analytic efforts (for example, classification of key themes or stylistic tones). Here, we leveraged contextual analysis to find relevant search phrases. Around each target word, the tool provides the words to the left and right by *X* number of words (say, five words to the left and five to the right). (Note that the keyword *decision* was not listed above but is an example of an expanded search word used during the course of the research.)

The search procedure allowed the research team to determine a subset of PEs associated with data analytics supporting acquisition decisionmaking. These PEs were then assessed across three subjective, mutually exclusive categories by the estimated fraction of dollars associated with data analytics: (1) all or nearly all, (2) moderate, and (3) small part of total.

PE Counts (FY 2019 PB for RDT&E)

	Budget Activity							
Organization	**01**	**02**	**03**	**04**	**05**	**06**	**07**	**Row Total**
AF	3	12	14	41	60	32	147	309
ARMY	4	27	24	30	68	29	52	234
NAVY	3	14	13	73	80	26	55	264
OSD	3	6	24	15	11	25	7	91
All Others (MDA, DTRA, SOCOM, etc.)	4	24	21	35	8	27	44	163
Column Total	**17**	**83**	**96**	**194**	**227**	**139**	**305**	**1061**

SOURCES: ASAF(FMC), 2018a, 2018b, 2018c, 2018d; ASA(FMC), 2018a, 2018b, 2018c, 2018d, 2018e, 2018f, 2018g; ASN(FMC), 2018a, 2018b, 2018c, 2018d, 2018e; USD(Comptroller) 2018a, 2018b, 2018c, 2018d, 2018e, 2018f, 2018g, 2018h.

NOTE: DTRA = Defense Threat Reduction Agency; SOCOM = Special Operations Command; ASAF(FMC) = Assistant Secretary of the Air Force for Financial Management and Comptroller; ASA(FMC) = Assistant Secretary of the Army for Financial Management and Comptroller; ASN(FMC) = Assistant Secretary of the Navy for Financial Management and Comptroller.

The number of PEs in the FY 2019 PB for RDT&E is 1,061, as shown in the slide above. The RDT&E volumes used are included in multiple references (ASAF[FMC], 2018a, 2018b, 2018c, 2018d; ASA[FMC], 2018a, 2018b, 2018c, 2018d, 2018e, 2018f, 2018g; ASN[FMC], 2018a, 2018b, 2018c, 2018d, 2018e; USD[Comptroller], 2018a, 2018b, 2018c, 2018d, 2018e, 2018f, 2018g, 2018h).

31 PEs (~3% of 1,061 PEs) have indications of analysis supporting acquisition decision-making and management

Organization	BA	Budget Activity Title	PE	Program Element Title	FY 2019 Total $ (000s)
Air Force	01	Basic Research	0601102F	Defense Research Sciences	348,322
Air Force	02	Applied Research	0602201F	Aerospace Vehicle Technologies	130,547
Air Force	06	Management Support	0605101F	RAND Project Air Force	34,614
Air Force	06	Management Support	0702806F	Acquisition and Management Support	12,367
Air Force	06	Management Support	0605898F	Management HQ - R&D	10,642
Air Force	07	Operational Systems Development	0304310F	Commercial Economic Analysis	3,472
Air Force	06	Management Support	0804731F	General Skill Training	1,448
Army	05	System Development & Demonstration	0605013A	Information Technology Development	113,758
Army	01	Basic Research	0601104A	University and Industry Research Centers	92,115
Army	06	RDT&E Management Support	0605803A	Technical Information Activities	29,050
Army	03	Advanced Technology Development	0603734A	Military Engineering Advanced Technology	25,864
Army	06	RDT&E Management Support	0605103A	Rand Arroyo Center	19,821
DARPA	01	Basic Research	0601101E	Defense Research Sciences	422,130
DCMA	05	System Development And Demonstration	0605013BL	Information Technology Development	11,988
DLA	03	Advanced Technology Development	0603712S	Generic Logistics R&D Technology Demonstrations	11,778
DTIC	06	Management Support	0605801KA	Defense Technical Information Center	56,853
Joint Staff	06	Management Support	0204571J	Joint Staff Analytical Support	6,658
Navy	01	Basic Research	0601153N	Defense Research Sciences	458,708
Navy	02	Applied Research	0602750N	Future Naval Capabilities Applied Research	147,771
Navy	06	Management Support	0605154N	Center for Naval Analyses	48,797
Navy	06	Management Support	0605865N	Operational Test and Evaluation Capability	21,554
OSD	03	Advanced Technology Development	0603826D8Z	Quick Reaction Special Projects	69,626
OSD	01	Basic Research	0601110D8Z	Basic Research Initiatives	42,702
OSD	06	Management Support	0605104D8Z	Technical Studies, Support and Analysis	22,576
OSD	03	Advanced Technology Development	0603288D8Z	Analytic Assessments	19,472
OSD	03	Advanced Technology Development	0603833D8Z	Engineering Science & Technology	19,415
OSD	03	Advanced Technology Development	0603781D8Z	Software Engineering Institute	15,050
OSD	02	Applied Research	0602751D8Z	Software Engineering Institute (SEI) Applied Research	9,300
OSD	07	Operational System Development	0307577D8Z	Intelligence Mission Data (IMD)	6,889
OSD	06	Management Support	0606100D8Z	Budget and Program Assessments	5,768
OSD	04	Advanced Component Development And Prototypes	0603821D8Z	Acquisition Enterprise Data & Information Services	2,506

SOURCES: Analysis of ASAF(FMC), 2018a, 2018b, 2018c, 2018d; ASA(FMC), 2018a, 2018b, 2018c, 2018d, 2018e, 2018f, 2018g; ASN(FMC), 2018a, 2018b, 2018c, 2018d, 2018e; USD(Comptroller) 2018a, 2018b, 2018c, 2018d, 2018e, 2018f, 2018g, 2018h.

RAND CORPORATION

Insight from the RDT&E Budget Assessment

The 31 PEs (approximately 3 percent of the 1,061 PEs) that the research team determined have indications of analysis supporting acquisition decisionmaking are presented in the slide above (ASAF[FMC], 2018a, 2018b, 2018c, 2018d; ASA[FMC], 2018a, 2018b, 2018c, 2018d, 2018e, 2018f, 2018g; ASN[FMC], 2018a, 2018b, 2018c, 2018d, 2018e; USD[Comptroller], 2018a, 2018b, 2018c, 2018d, 2018e, 2018f, 2018g, 2018h). The associated component organization sponsoring the research, budget activity, program element number, and title, as well as the total project dollars, are provided. The table is sorted by sponsor organization and budget activity. We provide the number of PEs associated with each sponsor and each budget activity:

Breakout by Budget Activity (BA)	**Breakout by Sponsor**
BA1: 5	Air Force: 7
BA2: 3	Army: 5
BA3: 6	DARPA, DCMA, Defense Logistics Agency, Defense Technical Information Center, Joint Staff: 1 each
BA4: 1	Navy: 4
BA5: 2	OSD: 10
BA6: 12	—
BA7: 2	—

Additionally, the identified PEs' budget total ranged from the $458 million for the PE "defense research sciences" to $1.4 million for the PE "general skill training."

Having identified these candidate PEs, the research team next conducted a detailed examination of the justification descriptions to ascertain to what extent the analytics in question are specifically explained. The approach is subjective, and the justification exhibits, themselves, do not necessarily capture all supporting activities in detail. As an example, PE 0601153N, "defense research sciences," represents the U.S. Navy's S&T investment and supports a variety of basic research activities. The exhibit R-2 specifically states that, "due to the number of efforts in this PE, the programs described herein are representative of the work included in this PE" (ASN[FMC], 2018a).

We assessed how much of each PE appears relevant to acquisition data analytics

Organization	PE	Program Element Title	FY 2019 Total $ (000s)	*All or Nearly All of Total*	*Moderate*	*Small Part of Total*
Air Force	0601102F	Defense Research Sciences	348,322			X
Air Force	0602201F	Aerospace Vehicle Technologies	130,547			X
Air Force	0605101F	RAND Project Air Force	34,614		X	
Air Force	0702806F	Acquisition and Management Support	12,367		X	
Air Force	0605898F	Management HQ - R&D	10,642			X
Air Force	0304310F	Commercial Economic Analysis	3,472	X		
Air Force	0804731F	General Skill Training	1,448			X
Army	0605013A	Information Technology Development	113,758			X
Army	0601104A	University and Industry Research Centers	92,115			X
Army	0605803A	Technical Information Activities	29,050		X	
Army	0603734A	Military Engineering Advanced Technology	25,864		X	
Army	0605103A	Rand Arroyo Center	19,821		X	
DARPA	0601101E	Defense Research Sciences	422,130			X
DCMA	0605013BL	Information Technology Development	11,988			X
DLA	0603712S	Generic Logistics R&D Technology Demonstrations	11,778			X
DTIC	0605801KA	Defense Technical Information Center	56,853		X	
Joint Staff	0204571J	Joint Staff Analytical Support	6,658		X	
Navy	0601153N	Defense Research Sciences	458,708			X
Navy	0602750N	Future Naval Capabilities Applied Research	147,771			X
Navy	0605154N	Center for Naval Analyses	48,797		X	
Navy	0605865N	Operational Test and Evaluation Capability	21,554			X
OSD	0603826D8Z	Quick Reaction Special Projects	69,626			X
OSD	0601110D8Z	Basic Research Initiatives	42,702			X
OSD	0605104D8Z	Technical Studies, Support and Analysis	22,576		X	
OSD	0603288D8Z	Analytic Assessments	19,472		X	
OSD	0603833D8Z	Engineering Science & Technology	19,415		X	
OSD	0603781D8Z	Software Engineering Institute	15,050		X	
OSD	0602751D8Z	Software Engineering Institute (SEI) Applied Research	9,300		X	
OSD	0307577D8Z	Intelligence Mission Data (IMD)	6,889			X
OSD	0606100D8Z	Budget and Program Assessments	5,768		X	
OSD	0603821D8Z	Acquisition Enterprise Data & Information Services	2,506	X		

SOURCES: Analysis of ASAF(FMC), 2018a, 2018b, 2018c, 2018d; ASA(FMC), 2018a, 2018b, 2018c, 2018d, 2018e, 2018f, 2018g; ASN(FMC), 2018a, 2018b, 2018c, 2018d, 2018e; USD(Comptroller) 2018a, 2018b, 2018c, 2018d, 2018e, 2018f, 2018g, 2018h.

RAND CORPORATION

After identifying the candidate PEs that appear to have analytic activities supporting acquisition decisionmaking, the research team sought to classify each PE in terms of how much of the total budget was dedicated to those analytic activities. The slide above provides the assessment of the 31 PEs regarding the apparent relevance of the work to defense acquisition analysis across the three subjective categories (all or nearly all of total, moderate, and small part of total) (ASAF[FMC], 2018a, 2018b, 2018c, 2018d; ASA[FMC], 2018a, 2018b, 2018c, 2018d, 2018e, 2018f, 2018g; ASN[FMC], 2018a, 2018b, 2018c, 2018d, 2018e; USD[Comptroller], 2018a, 2018b, 2018c, 2018d, 2018e, 2018f, 2018g, 2018h).

We next provide an example from each of the three assessment categories.

Example 1: All or nearly all of requested dollars appear relevant for PE 0603821D8Z, Acquisition Enterprise Data & Information Services

UNCLASSIFIED

Exhibit R-2, RDT&E Budget Item Justification: PB 2019 Office of the Secretary Of Defense | **Date:** February 2018

Appropriation/Budget Activity
0400: *Research, Development, Test & Evaluation, Defense-Wide* / BA 4: *Advanced Component Development & Prototypes (ACD&P)*

R-1 Program Element (Number/Name)
PE 0603821D8Z / *Acquisition Enterprise Data & Information Services*

COST ($ in Millions)	Prior Years	FY 2017	FY 2018	FY 2019 Base	FY 2019 OCO	FY 2019 Total	FY 2020	FY 2021	FY 2022	FY 2023	Cost To Complete	Total Cost
Total Program Element	0.000	1.761	2.198	2.506	-	2.506	3.071	3.925	3.987	4.060	Continuing	Continuing
840: *Acquisition Enterprise Data & Information Services*	0.000	1.761	2.198	2.506	-	2.506	3.071	3.925	3.987	4.060	Continuing	Continuing
Quantity of RDT&E Articles	-	-	-	-	-	-	-	-	-	-		

A. Mission Description and Budget Item Justification

The Acquisition Enterprise Data & Information Services (AEDIS) investment supports enhanced Acquisition Visibility (AV) for the Defense Acquisition Executive (DAE), Component Acquisition Executives (CAE), Service Chiefs of Staff, OSD senior leaders, and OSD and Component analysts who assess and decide the efficiency and effectiveness of acquiring and sustaining the Department's acquisition programs including Major Defense Acquisition Programs (MDAPs), Major IT investments, and Acquisition Category (ACAT) II – IV programs. AEDIS/AV supports DAE, GAE, and Service Chief responsibilities by providing critical information for acquisition analysis, oversight, and decisions. AEDIS/AV institutionalizes the management of data and business rules used in the Department's acquisition decision making, and it integrates the acquisition data stored across multiple disparate Federal and Departmental organizations' data sets and systems. The AEDIS/AV investment delivers a Department-wide accessible collection of acquisition information, techniques, and tools, including the Defense Acquisition Visibility Environment (DAVE), the Defense Acquisition Management Information Retrieval (DAMIR) capability, and acquisition data analysis capabilities as well as data access services and data standards via the Acquisition Visibility Data Matrix (AVDM). Funding supports enhancements to Acquisition Visibility through the definition, development, and fielding of concepts and tools for Department-wide data analysis for use across Congress and the Department, particularly in support of the DAE and his decision authority.

B. Program Change Summary ($ in Millions)	FY 2017	FY 2018	FY 2019 Base	FY 2019 OCO	FY 2019 Total
Previous President's Budget	2.136	2.198	2.523	-	2.523
Current President's Budget	1.761	2.198	2.506	-	2.506
Total Adjustments	-0.375	0.000	-0.017	-	-0.017
• Congressional General Reductions	-	-			

RAND

SOURCE: USD(Comptroller) (2018e).

The slide above provides an example assessment for "all or nearly all": PE 0603821D8Z, "acquisition enterprise data & information services." The mission description for this PE indicates that the investment supports "Acquisition Visibility (AV) for the Defense Acquisition Executive (DAE), Component Acquisition Executives (CAE), Service Chiefs of Staff, OSD senior leaders, and OSD and Component analysts who assess and decide the efficiency and effectiveness of acquiring and sustaining the Department's acquisition programs" (USD[Comptroller], 2018e). An example of a specific capability is enhancements to "Acquisition Visibility through the definition, development, and fielding of concepts and tools for Department-wide data analysis for use across Congress and the Department, particularly in support of the DAE and his decision authority" (USD[Comptroller], 2018e). Specific language that assisted with the assessment is highlighted in yellow.

Example 2: Small amount of total requested dollars appear relevant for PE 0603288D8Z, Science & Technology Analytic Assessments

UNCLASSIFIED

Exhibit R-2A, RDT&E Project Justification: PB 2019 Office of the Secretary Of Defense		Date: February 2018
Appropriation/Budget Activity 0400 / 3	**R-1 Program Element (Number/Name)** PE 0603288D8Z / *Science and Technology (S&T) Analytic Assessments*	**Project (Number/Name)** 328 / *Science and Technology Analytic Assessments*

B. Accomplishments/Planned Programs ($ in Millions)	FY 2017	FY 2018	FY 2019
following capability areas: foreign, integrated air and missile defense capabilities; options for U.S. electronic warfare and capability to counter adversaries; resiliency in U.S. Command, Control, Communications, Computers, Intelligence, Surveillance, and Reconnaissance (C4ISR) systems and options to counter adversaries C4ISR capabilities; ground combat offensive and defensive capabilities, air dominance and missile defense, and undersea engagements. The QRAT is enabled by a weekly meeting of FFRDC/UARC lead contacts to review on-going and emerging tasks and collaborative technical interchanges on OUSD(R&E) and OUSD(A&S) focus areas. Technical Analysis (Strategic Studies): Strategic studies are 6-12 month engineering level systems analysis. Strategic studies parametrically define the emerging threat space, determine feasibility of potential solutions and parametrically analyze the solution trade space. Specific tasks that will be executed within the strategic studies area include: - Evaluate options to counter foreign missile capabilities. - Explore feasibility and potential of next generation electronic warfare technologies. - Characterize an architecture for theater-level electronic warfare threat awareness and battle management to effectively and efficiently apportion resource in a constrained environment. - Identify future threat detection and identification capabilities for future electronic support systems. - Evaluate threats to High Value Air Assets (HVAA) and identify potential countermeasures to develop and alternative ways to accomplish the HVAA missions. - System and technology assessments for surface and sub-surface warfare. - Evaluate options for land based defense against a missile raid. - Evaluate options for maritime based defense against a missile raid. - Evaluate efficacy of passive systems and counters to passive systems. Analytic Tools: - Develop analytic tools to inform and evaluate new technologies' potential to counter emerging threats and exploit adversary vulnerabilities from air, land, sea, and space domains. - Develop of analytic tools to provide inform and provide decision support to resourcing recommendations. - Develop integrated modeling, simulation, and analysis tools to aid complex acquisition decisions. - Develop Red Teaming methodology for evaluating US capabilities and systems in the context of emerging threats in relevant scenarios.			

RAND CORPORATION

SOURCE: USD(Comptroller) (2018d).

The slide above provides an example assessment for "small part": PE 0603288D8Z, "science & technology analytic assessments" (USD[Comptroller], 2018d). The mission description for this PE indicates, "This Program Element (PE) directly supports the Office of the Under Secretary of Defense, Research and Engineering (OUSD[R&E]) and the OUSD(A&S) with assessments and analysis to inform the strategic direction of research, development, and acquisition of innovative capabilities" (USD[Comptroller], 2018d). This justification also contained one of the few instances of a proportional division of the supported activities: "Typically, the ratios of resources applied to Operational and Technical Assessments, Technical Analysis and Quick Reaction Analysis Team, and development of Analytic Tools will be roughly 30/60/10 percent" (USD[Comptroller], 2018d). Specific language that assisted with the assessment is highlighted in yellow.

Example 3: RAND Project Air Force, PE 0605101F, is analysis focused, but not all of the work performed is related to acquisition

UNCLASSIFIED

Exhibit R-2, RDT&E Budget Item Justification: PB 2019 Air Force — **Date:** February 2018

Appropriation/Budget Activity
3600: *Research, Development, Test & Evaluation, Air Force* / BA 6: *RDT&E Management Support*

R-1 Program Element (Number/Name)
PE 0605101F / *RAND Project Air Force*

COST ($ in Millions)	Prior Years	FY 2017	FY 2018	FY 2019 Base	FY 2019 OCO	FY 2019 Total	FY 2020	FY 2021	FY 2022	FY 2023	Cost To Complete	Total Cost
Total Program Element	-	33.373	34.346	34.614	0.000	34.614	35.258	35.869	36.614	37.282	Continuing	Continuing
661110: *Project Air Force*	-	33.373	34.346	34.614	0.000	34.614	35.258	35.869	36.614	37.282	Continuing	Continuing
Quantity of RDT&E Articles	-	-	-	-	-	-	-	-	-	-		

A. Mission Description and Budget Item Justification

This program provides for continuing analytical research across a broad spectrum of aerospace issues and concerns. The Project AIR FORCE (PAF) research agenda is focused primarily on mid to long-term problems; in addition, PAF provides quick response assistance for senior Air Force officials on high priority, near term issues. Within these areas, PAF addresses difficult and complex, far-reaching and inter-related questions linked to future strategies, approaches and policies, in order to enhance Air Force senior leadership's deliberations and decisionmaking on major issues. The Air Force Steering Group, chaired by the Vice Chief of Staff, reviews, monitors, and approves PAF annual research efforts. Each project is initiated, processed, and approved in accordance with PAF Sponsoring Agreement which requires General Officer (or SES equivalent) sponsorship and involvement on a continuing basis.

PAF is organized in four primary research program areas: strategy and doctrine; force modernization employment; manpower, personnel and training; and resource management. Integrative research projects are also conducted at the division level with direct support provided through the most applicable program. Research programs address organizational crosscutting issues as defined by specific research themes approved by the Air Force Steering Group. These research themes encompass a wide spectrum of topics including external challenges to national security; terrorism and homeland defense; joint and coalition operations; integrated roadmap for ISR capabilities; enhancing, tailoring and reducing infrastructure to meet new force requirements; potential changes to the Active/Reserve/National Guard/Civilian/Contractor manpower mix; and improved weapon system costing.

The research program will continue to build upon research foundations, examining the evolving security environment, emerging threats, national and military strategy, transformation approaches including investment strategies to provide capabilities within changing Defense budgets, operational concepts to meet evolving and increasingly joint missions, exploiting advanced technologies, increasing the effectiveness and efficiency of combat support, and developing the total force (Active/Reserve/National Guard/Civilian/Contractor). These efforts will continue to inform and support the senior Air Force leadership regarding personnel management and training; improving logistical efficiencies and force sustainment; ongoing conflicts and joint operations; force structure capabilities, limitations, and operational concepts; and making force structure tradeoffs within resource constraints to meet future national security and Air Force needs.

Future research will build upon earlier work to continue to help the Air Force to rapidly and appropriately adapt to the changing world environment and emerging threats; continue to modernize and employ its force structure to provide capabilities within changing DoD budgets; assess lessons learned from recent and ongoing conflicts; develop and utilize its total force; and enhance the support of our aerospace forces, ranging from sustainment of the force structure to agile combat support.

PAF research spans functional and organizational boundaries and is managed in a manner to facilitate independence and freedom from organizational bias thereby providing perspectives and insights to senior Air Force leaders free from parochial influences not necessarily in the best interest of the Air Force at large. Benefits of

RAND

SOURCE: ASAF(FMC), 2018b.

The table shown in the slide above provides an example assessment for "moderate": PE 0605101F, "RAND Project Air Force."[10] The mission description for this PE indicates: "This program provides for continuing analytical research across a broad spectrum of aerospace issues and concerns" (ASAF[FMC], 2018b). The justification description also indicates that "the efforts will continue to inform and support the senior Air Force leadership regarding . . . improving logistical efficiencies and force sustainment . . . and making force structure tradeoffs within resource constraints to meet future national security and Air Force needs." The research team's knowledge of this PE allowed us to extend this same assessment category to other FFRDC-related PEs. Specific language that assisted with the assessment is highlighted in yellow.

[10] Disclosure: Some of the authors perform research within RAND Project Air Force. This example was included because it is one of the few FFRDCs that has a line item in the DoD's budget and because we could employ our personal insight into the activities of RAND Project Air Force to help assess the fidelity and clarity of the budget description against data analytics activities.

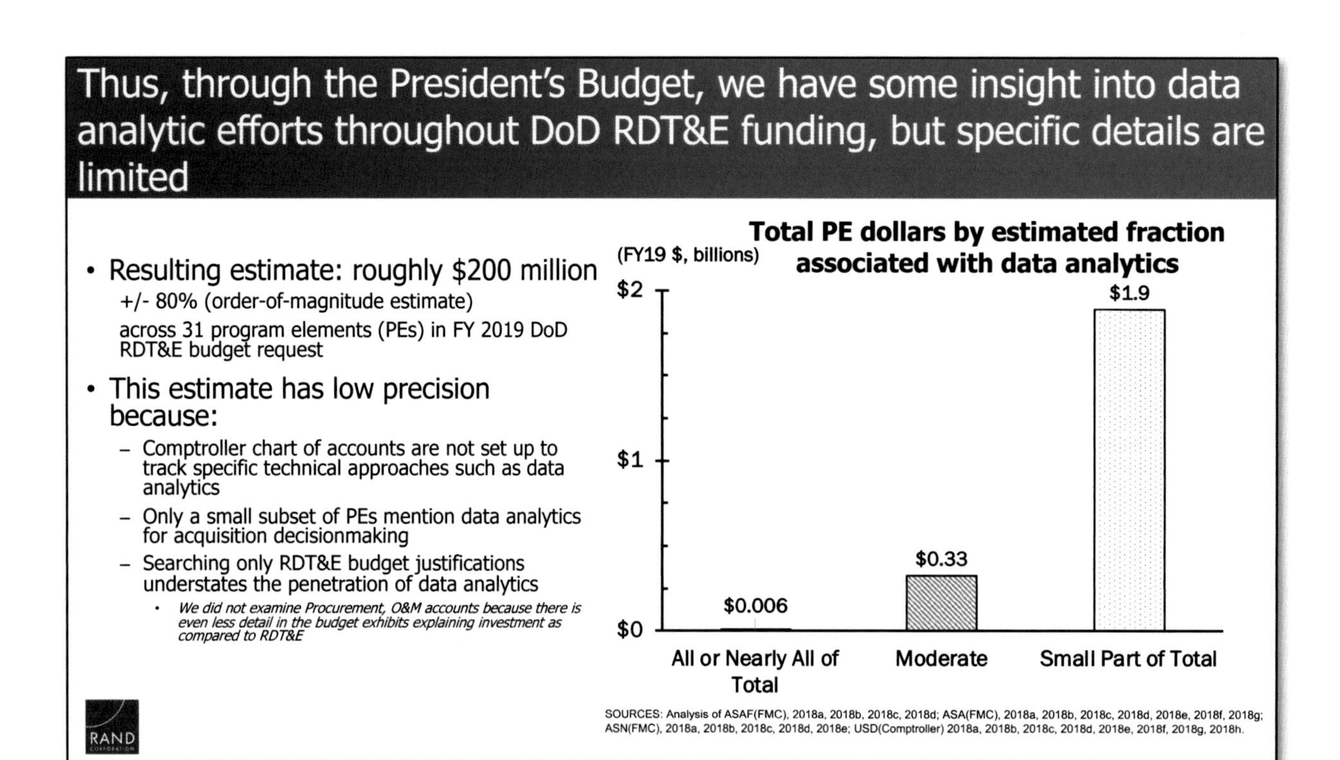

The slide above presents the individual assessments categories as a summary graph. The two PEs assessed as "all or nearly all" total about $6 million. Fourteen PEs assessed as "moderate" total about $330 million. Fifteen PEs totaling $1.9 billion were assessed as "small part of total" funding associated with data analytics supporting acquisition decisionmaking (ASAF[FMC], 2018a, 2018b, 2018c, 2018d; ASA[FMC], 2018a, 2018b, 2018c, 2018d, 2018e, 2018f, 2018g; ASN[FMC], 2018a, 2018b, 2018c, 2018d, 2018e; USD[Comptroller], 2018a, 2018b, 2018c, 2018d, 2018e, 2018f, 2018g, 2018h). Given those rough assessments, the total investments may be about $200 million for FY 2019 (plus or minus roughly 80 percent); however, that is a rough-order-of-magnitude approximation, given the fidelity of the qualitative descriptions. Still, this gives an indication that there are activities across the DoD with a rough sense of its magnitude.

It is important to note several caveats to this analytic effort. First, the comptroller chart of accounts is not set up to track specific technical approaches, such as data analytics, although the justification descriptions provide the opportunity for highlighting such efforts. Second, only a small subset of PEs mention data analytics for acquisition decisionmaking. It is unknown whether this information truly indicates how few PEs are using or developing data analytics or reflects the nature and requirements of budget justification statements. Finally, we did not examine procurement and operations and maintenance accounts because the budget exhibits include even less detail that explains investment than do RDT&E exhibits. Thus, at the PB level, we have some insight into data analytics efforts throughout DoD RDT&E funding, but specific details are limited.

What's in this Briefing

- Background
- Approach and methodology
- Findings on congressional questions
  - Q1–Q2: Extent of, and potential to increase, DoD data analytics capabilities
  - Q3: R&D funding for data analytics capabilities
    - *Program-element RDT&E budgets*
    - *IT system RDT&E budgets*
    - *SBIR and STTR solicitations*
  - Q4: Private-sector best practices for efficient data collection and delivery
  - Q5: Anonymized data sharing with researchers and analysts
  - Q6: Appropriate data analytics and methods courses at defense training institutions
- Conclusions

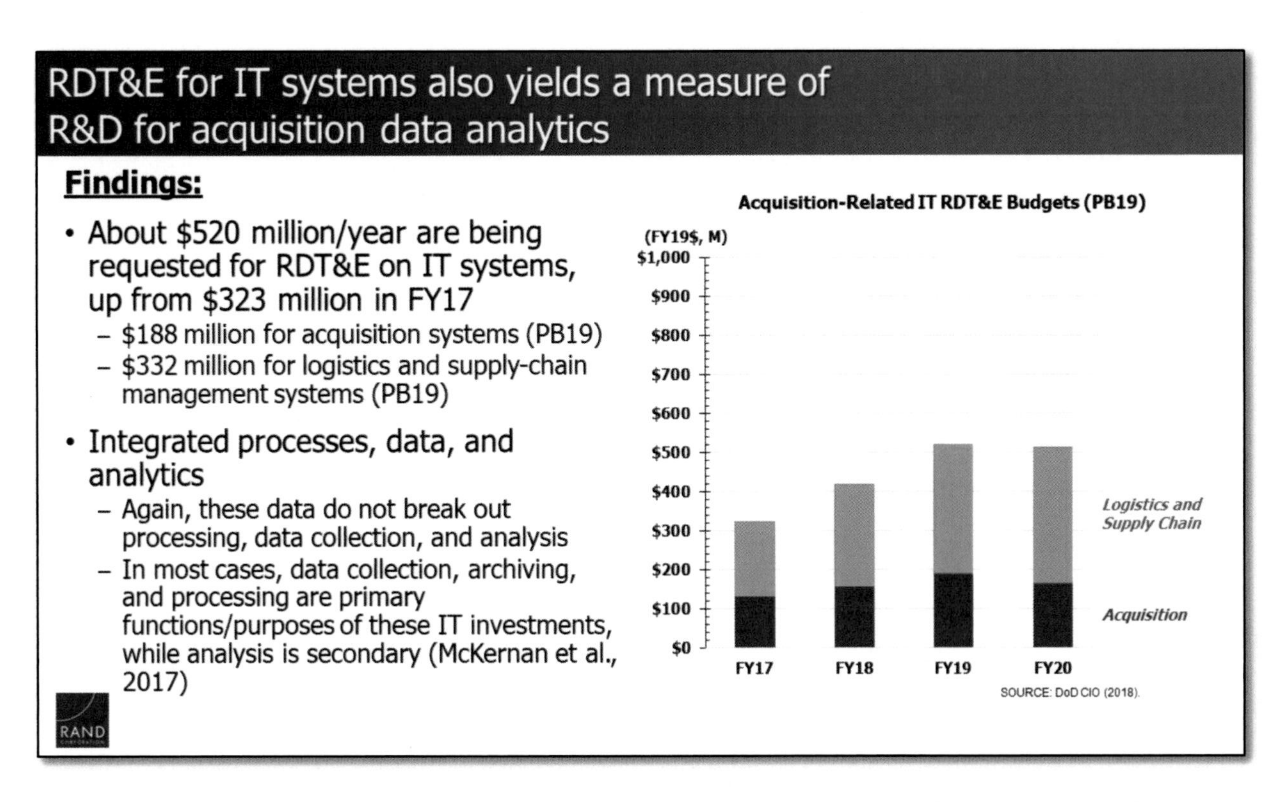

For question 3, we also examined the RDT&E budgets associated with IT systems in SNaP-IT budget exhibits for PB 2019 (DoD CIO, 2018). Those budget request details indicate both the topic areas (acquisition or logistics and supply-chain management) and the funding type

(RDT&E, in this case). The requests for FY 2019 and FY 2020 were $520 million and $514 million, respectively (in adjusted FY 2019 dollars). Those investments represent an increase from the $323 million and $417 million (adjusted FY 2019 dollars) reported for FY 2017 and FY 2018, respectively.

As with the earlier analysis of total IT investments (not just RDT&E), these data give a rough measure of the scope of information system investments but have some limitations. First, even though the broad functional areas are identified in SNaP-IT, it is difficult to assess precisely what each IT system does in detail across all the systems listed. The split among business processes, information collection, and analysis is not clearly delineated. Each investment has a short description, but it does not provide a structured breakdown. Prior research by McKernan et al. (2017) indicates that most of the major systems examined are dominated by process support, data collection, and data archiving, with analytic layers being secondary.

SBIR and STTR Solicitations

Some advanced analytics are being pursued via SBIR and STTR for design tools and production improvements

Selected DoD SBIR 19.1 and STTR 19.A Topics (issued 1/8/19; closed 2/6/19)

- Design and processing
  - Data Analytics and Machine Learning to Accelerate Materials Design and Processing *(Navy N19A-T020)*
  - Advanced Comprehensive Analysis Tool for Emerging Vertical Lift Configurations *(Army A19-001)*
- General
  - Machine Learning Dataset Auto Generator *(Army A19-028)*
  - Autonomous Decision Making via Hierarchical Brain Emulation *(Air Force AF19A-T009)*

Finally, we examined the recent SBIR and STTR solicitations (19.1 and 19.A, respectively) that were issued on January 8, 2019, and closed on February 6, 2019. SBIR is a competitive extramural R&D funding activity for domestic small businesses on topics of relevance to the issuing agency that have the potential for commercialization. STTR provides extramural R&D funding to support joint ventures between small businesses and nonprofit research institutions to facilitate technology transfer (see U.S. Small Business Administration, undated). We note that SBIR and STTR solicitations might not result in awards, and any such awards are relatively small, compared with PEs, but we include them to try to measure DoD interest in R&D for acquisition data analytics.

This examination identified the R&D topics listed in the slide above that are related to data analytics. These topics are grouped into two general categories: (1) system design and processing and (2) general data analytics. Although these topics are solicitations and have not yet become awards, they indicate the types of data analytics investments sought to improve defense acquisition.

There were five other solicitations involving analytic techniques for incorporation in the systems to be acquired, but we did not list these, since this study did not include operational uses of data analytics (including uses in the acquired systems that support operations). A review of such uses would require a much broader examination.

Question 4. Private-Sector Best Practices for Minimizing the Collection and Delivery of Data

What's in this Briefing

- Background
- Approach and methodology
- Findings on congressional questions
  - Q1–Q2: Extent of, and potential to increase, DoD data analytics capabilities
  - Q3: R&D funding for data analytics capabilities
  - Q4: Private-sector best practices for efficient data collection and delivery
  - Q5: Anonymized data sharing with researchers and analysts
  - Q6: Appropriate data analytics and methods courses at defense training institutions
- Conclusions

Congress's fourth topic area asks about "any potential improvements, based on private-sector best practices, in the efficiency of current data collection and analysis processes that could minimize collection and delivery of data by, from, and to government organizations" (U.S. House of Representatives, 2016, p. 1125).

DoD uses most private-sector best practices for efficient data collection and delivery, but data sharing remains a challenge

- Private-Sector best practices include:
  - Identify specific questions that leadership needs to answer in order to make informed decisions
  - Plan and prioritize what data are needed using an organization's data strategy
  - Define data and establish common definitions between organizations using that data
  - Designate single authoritative source for each datum or dataset, then share existing authoritative datasets between systems via technical means
  - Emphasize that data are corporate-wide assets, not owned by local units

- Current status in DoD:
  - DoD is pursuing most of these practices
    - Common PM tool suites are less common but being pursued
    - Other acquisition IT systems provide data feeds (e.g., contracting, auditing)
      - This is measured in the SNaP-IT data presented earlier
    - Open systems and data sharing have reduced duplication and ensured common data
  - Data sharing and ownership challenges remain
    - Security, trust, and sensitivity meta-labeling
  - Opportunities to improve include:
    - Continued pursuit of these practices
    - Address disincentives to data sharing
      - Clarify oversight extent and roles
        - Nominal by ACAT level; elevate when problems occur
      - Pursue data-element sensitivity meta-labeling
      - Understand sensitivity upgrades from data-aggregation

NOTE: ACAT = acquisition category.

Best Practices

A number of private-sector best practices (identified through literature reviews, prior research, and attendance at domain meetings) could improve DoD efficiency by minimizing the collection and delivery of data.[11] One identified best practice is to objectively plan and prioritize what data are essential to collect in the first place (i.e., define the organization's data strategy) by identifying specific questions that leadership needs answered to make informed decisions, the essential data options needed for analysis, and the costs and benefits of these alternatives (Nott, 2015; *MIT Sloan Management Review*, 2016; vander Meulen and McCall, 2018; DalleMule and Davenport, 2017). One specific approach is to identify and prioritize such use cases as vehicles for planning and management (Fleming et al., 2018).

A second set of practices is defining the data, establishing common definitions between organizations using those data, and automatically collecting data from operational systems for subsequent analysis and use by management and oversight (Taylor, 2016). Automatically collecting data alleviates manual data reporting and can provide more-accurate, -current, and -detailed data on activities and functions. Transmittal of such data is performed seamlessly by interconnected major IT systems.

[11] These are best practices in that they are from consulting companies that assess, survey, and review the field for lessons learned and common practices. We also found them to be relatively consistent.

A third practice is to designate which data system will be the single authoritative source for a datum (DalleMule and Davenport, 2017), then share it via technical means to other systems that need the datum. This increases transparency and ensures that everyone is using the same data while reducing duplicative, and potentially erroneous, data entry. Finally (and perhaps most important), the private sector emphasizes that data are corporate-wide assets. They are not owned and controlled by local units but are generally available widely (with appropriate privacy protections) to improve the efficiency of the organization (Svensson, 2013).

We assessed the DoD's status against these practices by comparing the DoD's status (as identified in prior research [Riposo et al., 2015; McKernan et al., 2016, 2017, 2018] and recent DoD information management meetings) against these best practices. Comparison is straightforward, seeing whether key practices are being used by the DoD.

Status in the DoD

The data managers across the DoD are pursuing many of these practices. Use cases are used by information managers to identify and manage key data. Designating authoritative sources and sharing data among systems have become the standard approach by information managers, and that approach is maturing in many systems across acquisition functions. Common PM software suites that can automatically share data, such as program cost, schedule, and status information, are less common but are reflected in current DoD plans. However, the broader extent of IT systems measured earlier using SNaP-IT data are generally such systems; they support direct operations while recording and sharing data for analysis (see earlier analysis on Congress's questions 1 and 2). Such considerations as ensuring that draft data are not transmitted until they are finalized will have to be addressed when selecting or designing software suites.

Although the DoD has made progress in opening and sharing its acquisition data systems, challenges remain. The general culture against sharing is driven by various disincentives and concerns, including security (e.g., sensitivity elevation from data aggregation), trust (e.g., concern that oversight support staff will use data to micromanage projects or that authorized users might inappropriately release data), and data labeling (e.g., a general lack of metadata on the sensitivity level of individual data elements).

Note that in the private sector, the value that data bring to the organization (e.g., increased sales and profit) is easy to calculate, but in the DoD it is more difficult to tie data to value. The DoD collects data for many reasons that are not necessarily directly tied to helping the warfighter, including auditing and other external inquiries. These objectives need to be factored into the data management plan.

Opportunities

Opportunities for the DoD include continued pursuit of these best practices, addressing the disincentives to share, and developing approaches to resolve security and sensitivity issues.

For several years, DoD has been implementing a number of best practices to increase the efficiency of data collection, delivery, and use

- Strategic level:
  - DoD leadership support for data-driven decision-making
  - Supported and used analytic organizations
    - *e.g., CAPE and OSD Acquisition Analytics (previously PARCA)), FFRDCs, and UARCs*
  - Continued to advance and coordinate data systems across DoD
    - *e.g., program information, contracting, contract management, payments, auditing*
  - Established the OSD CDO
- Operational level:
  - DoD acquisition information managers are employing a number of industry best practices
    - *Data management*
      - Determination/agreement on authoritative sources
      - Data definitions to facilitate common sharing and understanding of data across organizations
      - Use cases to understand data needs and ensure data are collected to satisfy needs
      - Use of an official acquisition program list
      - Moving from unstructured to structured data for key data
      - Leveraging text and other analytic tools for larger unstructured data
      - Utilizing a working group in order to resolve data-related issues
    - *Data transparency and sharing between information systems*
    - *Adding COTS analytic and visualization layers to data systems*
      - e.g., for PMRT, CADE, DAVE, DIBNow
    - *Creating virtual environments in which COTS analysis tools can be used securely*
    - *Using general licenses for analysis over DoD data in the cloud (e.g., DAVE/AL)*
    - *Pursuing PM software that can automatically report program status data for oversight*

For several years, the DoD has been implementing best practices to increase the efficiency of data collection, delivery, and use. These changes have been occurring at both the strategic and operational levels. We will provide some examples here. These examples are illustrative and do not include all efforts currently in progress.

At the strategic level, the DoD leadership has been emphasizing the need to use data in decisionmaking, rather than simply to collect data without a clear purpose. With leadership support, the organizational culture has begun to change. In addition, from an organizational perspective, the Weapon Systems Acquisition Reform Act of 2009 (Pub. L. 111-23, 2009) stood up two organizations with largely analytical missions: CAPE and PARCA.[12] The new OUSD(A&S) continues to include an organization with a mission involving analytics (Acquisition Analytics). Through the establishment of organizations with analytics as a core mission, the DoD is emphasizing the importance of analytics in its broader mission. The DoD also steadily utilizes the analytic capabilities of the FFRDCs and UARCs and others in the private sector, as needed, to answer difficult questions. Finally, at the strategic level, a CDO now exists within OSD. The Air Force and Army also have similar positions. CDOs are a best practice in the private sector for data governance and management. The DoD needs to work through how this position can best support DoD acquisition's tough data governance and management challenges.

[12] The act created CAPE, whose origins came from the prior Office of Program Evaluation and Analysis.

At the operational level, clear evidence exists for multiple best practices in data management, including determining and agreeing on authoritative sources. For instance, the cost data collected by CAPE is an authoritative source. FPDS-NG is the authoritative source for contracts data. OSD has also been working with the military departments to establish some definitions for various data elements that are being used throughout acquisition. One such example is the current effort to establish definitions for "Middle Tier" acquisition data collection. In addition, the DoD acquisition program community is becoming aware of and documenting the various use cases for program data within DAVE. DAVE is also the source of a recent effort in which the military departments worked with OSD to establish an official acquisition program list for the department. Next, the acquisition community has moved from unstructured to structured data in the vast majority of its core information systems, so that it will be easier to use this information for decisionmaking, execution, and analysis. For data that remain unstructured (e.g., archiving of approved acquisition documentation), the DoD is leveraging text and other analytic tools. The program data information managers within OSD and the military departments, along with other relevant information managers, participate in the Acquisition Visibility Steering Group and Working Group to solve tough data-related problems.

Also, at the operational level, data transparency and sharing among information systems has greatly improved as the increasing or opening of APIs to data systems has been established. This best practice has been supported by the Open Data Policy, Digital Strategy, and U.S. CIO. This practice has improved the quality and efficiency of the data and collection. The DoD acquisition community has also added some COTS analytic and visualization layers to some of its newer information systems, including PMRT, CADE, DAVE, and DIBNow. These layers employ some of the latest private-sector software. Parts of the DoD are now using virtual computing environments in which COTS analysis tools can be used securely (alleviating the need to thoroughly test these tools for cybersecurity vulnerabilities before loading on DoD computers). Additionally, OSD(A&S/Acquisition Data) has added an analytic layer to DAVE that allows more advanced data scientists to utilize advanced software packages without the struggles of buying software or downloading it onto individual computers. Lastly, multiple efforts within the military departments are trying to identify and implement PM software that can automatically report program status data for oversight to some central information systems at the SAE level.

Question 5. Steps Being Taken to Expose Anonymized Data to Researchers and Analysts

What's in this Briefing

- Background
- Approach and methodology
- Findings on congressional questions
  - Q1–Q2: Extent of, and potential to increase, DoD data analytics capabilities
  - Q3: R&D funding for data analytics capabilities
  - Q4: Private-sector best practices for efficient data collection and delivery
  - Q5: Anonymized data sharing with researchers and analysts
  - Q6: Appropriate data analytics and methods courses at defense training institutions
- Conclusions

Congress's fifth question asked about which "steps are being taken to appropriately expose acquisition data in an anonymized fashion to researchers and analysts" (U.S. House of Representatives, 2016, p. 1125).[13]

[13] Some kind of data protection is often needed because acquisition data can be source-selection sensitive, classified, operationally sensitive, personally identifiable information (PII), predecisional, trade secrets, proprietary to contractors or commercial companies, restricted to the Committee on Foreign Investment in the United States, or other types of sensitivities.

DoD anonymization of acquisition data is minimal; anonymization has limited utility and DoD is pursuing other strategies to share data

- Generally, DoD is not anonymizing acquisition data for sharing
  - A counter-example is PII personnel data
- Anonymizing has limited utility
  - Some anonymization can be broken with modern correlation tools
  - Meta-data are required for correlation analysis and cannot be anonymized
  - Data sensitivity is often not stored at the data element level
  - Procedures for categorizing and handling sensitive data are complicated, slow, and not well understood; incentives impede progress
  - Aggregating data can increase data sensitivity and classification
- One possibility for future research might be blind analysis
  - Algorithms are designed and run in a black box, only showing the results
  - Challenges include data cleaning and understanding
- DoD is making other progress in sharing data
  - Identifying available data
  - Improving internal data transparency through technical means
  - Adding ability for researcher and analyst accounts on data systems
  - Removing classified and sensitive data elements before publishing (e.g., in FPDS-NG)

The balance between data sharing and privacy is of significant interest to governments, companies, and individuals. The benefits to data sharing are well documented, as is the need to protect the information in the process of sharing. For example:

> Proper deidentification enables organizations to safely share data sets for a broad range of valuable purposes without endangering the privacy interests of data subjects. . . . Government agencies routinely collect, process, and share huge troths of citizens' data for a wide range of administrative purposes and to ensure accountability regarding their own activities. Commercial firms providing financial, healthcare, retail or marketing services match or exceed government collection and use of data, and they often rely on deidentified data to develop or improve products and services. And, of course, academic researcher[s] rely on many sorts of data for a wide range of public health and social science research. More recently, and under the rubric of open data, governments and other large organizations have started to publicly release large data sets to promote the public good and lend support both to commercial endeavors and funded research. In short, deidentified data is a vital aspect of the digital economy. We all benefit from it in many ways ranging from education programs, to improved traffic flows and urban planning, to anti-theft and fraud programs, to genetic research. (Rubinstein, 2016, p. 2)

According to IBM, "Data anonymization helps you protect sensitive data, such as personally identifiable information or restricted business data to avoid the risk of compromising confidential information. It is defined in policy rules that are enforced for an asset. Depending on the method of data anonymization, data is redacted, masked, or substituted in the asset preview" (IBM Watson Studio, 2019).

Although the DoD has made some progress in improving data sharing, it is not generally anonymizing data as a means to further increase sharing, for various reasons discussed below.

One major exception is personnel data (including AWF data), which are often sensitive PII with legal releasability restrictions. Personnel data are often anonymized, but even such data are hard to get directly from the source (although some anonymized data can be obtained indirectly through OPM sources).

Although not widely using anonymization per se as a sharing technique, the DoD is increasing data availability by identifying available data, improving access through technical means and interfaces, adding the ability for some researchers and analysts to obtain accounts on some data systems, and transmitting unrestricted data after extracting sensitive and classified data (e.g., as is done with contracting data information sent to FPDS-NG).

Practical reasons explain why anonymization has not been widespread. Anonymization is not always reliable, and advances in analytic tools can sometimes identify data (e.g., in personnel data). For instance, in 2015, MIT researchers reported "that just four fairly vague pieces of information—the dates and locations of four purchases—are enough to identify 90 percent of the people in a data set recording three months of credit-card transactions by 1.1 million users" (Hardesty, 2015). The possibility that machines are able to identify sensitive DoD acquisition information despite applying anonymization techniques is a concern among the DoD's information managers.

Also, much of the metadata that would be removed in anonymization are important for trying to identify potential causes of identified trends. In addition, DoD data generally lack data sensitivity metadata at the data-element level, making it difficult to identify which data must not be shared and why. Furthermore, government procedures for categorizing and handling sensitive data are complicated, slow, and not well understood by staff, and incentives drive conservatism to block sharing (e.g., what exactly can and cannot be asserted as PROPIN by a contractor, how markings can be changed, and the personal risks involved).

One possible alternative to straightforward anonymization might be to allow researchers and analysts to develop algorithms that run on the raw data in a protected "black box" environment, exposing the results only to the researchers. This idea is just notional at this point, since issues of data cleaning, data understanding, and ensuring that the resulting analytic results do not divulge sensitive information would need to be resolved.

Question 6. Defense University Curricula on Data Analytics and Other Evaluation-Related Methods

What's in this Briefing

- Background
- Approach and methodology
- Findings on congressional questions
  - Q1–Q2: Extent of, and potential to increase, DoD data analytics capabilities
  - Q3: R&D funding for data analytics capabilities
  - Q4: Private-sector best practices for efficient data collection and delivery
  - Q5: Anonymized data sharing with researchers and analysts
  - Q6: Appropriate data analytics and methods courses at defense training institutions
- Conclusions

As the sixth and last question, Congress asked whether "the curriculum at the National Defense University [NDU], [DAU], and appropriate private-sector academic institutions include appropriate courses on data analytics and other evaluation-related methods and their application to defense acquisitions" (U.S. House of Representatives, 2016, p. 1125).

We reviewed the curricula at four defense institutions: NDU, DAU, Naval Postgraduate School (NPS), and the Air Force Institute of Technology (AFIT). We also examined acquisition-related offerings at five civilian universities: Georgia Institute of Technology (Georgia Tech), American University, George Mason University (GMU), Georgetown University, and George Washington University. Finally, we reviewed courses with the following strategic partners established by DAU: Johns Hopkins University, Stanford, University of Michigan, Google, IBM, and the DoD Cyber Crime Center (DC3) Cyber Training Academy. These courses are available for government and nongovernment employees in the DoD.

DoD's internal institutions have many applied data analytics and a few purely data science courses

- Defense Acquisition University (DAU)
  - Many applied data-analysis classes and a few general-purpose data-analytic courses
  - Other courses include methods directly tied to applications (e.g., PQM, testing, contracting)
  - Provides analytic "tools" (simpler calculation) in courses and online
- Naval Postgraduate School (NPS)
  - Offers a variety of data science classes teaching data methods, tools, and a variety of analysis
- Air Force Institute of Technology (AFIT)
  - Classes which use data are mostly applied data science courses teaching a variety of analysis. Some classes offered teaching methods and tools only
- National Defense University (NDU)
  - Most NDU courses are policy and strategy related rather than acquisition
  - Courses which incorporate data are mostly methods and tools, with some systems analysis
  - One "Big Data" course has been designed in response to the FY2017 NDAA

The DoD's defense universities have many data analytics courses, although most are applied rather than general data analytics courses. However, this observation is not unexpected, as the defense universities have a more tailored mission than does a general university. DAU has many applied data analysis classes (e.g., BCF 13, Applied Cost Analysis) and a few general-purpose data analytics courses (e.g., CMQ 131, Data Collection and Analysis). Other courses include methods directly tied to applications (e.g., PQM, testing, contracting), which is to be expected from an institution such as DAU. DAU also provides online analytic tools that are used in courses for anyone to use. Such tools include spreadsheet-based forms and other facilitators that guide the user into a data-driven methodology for a specific application. For example, a form coded in a spreadsheet based on the Federal Acquisition Regulations that prompts a contracting officer to determine a reasonable profit margin on a contract based on various data, such as contractor investments and infrastructure, is a simple applied tool that facilitates profit analysis based on data. Also, DAU tracks the views of its tools. Over the course of September 2018, the DAU tools website received more than 45,000 cumulative page views (DAU, personal communication).[14]

NPS offers a variety of data analytics classes that teach data methods, tools, and a variety of analyses (NPS, 2018, undated-a, undated-b). Although geared primarily toward Navy officers, the NPS courses are a good representation of data analytics courses.

---

[14] *Cumulative page views* are one type of data to help gauge use. However, this metric has some potential limitations (e.g., it might not reflect actual tool downloads or usage, it includes multiple views from the same user, and it can include errant navigation to the website).

AFIT has two primary schools. The School of Systems and Logistics offers acquisition courses and workshops that are largely management related (e.g., 21X 311, Depot Maintenance Operations), with relatively few data analytics courses. The School of Engineering and Management offers several acquisition-relevant programs (e.g., STAT 535, Applied Statistics for Managers II), all of which include data analytics courses. The School of Engineering and Management has several degrees and certificates that can be earned. Classes that use data are mostly applied data analytics courses that teach a variety of analyses. Some classes offer teaching methods and tools only.

Finally, NDU focuses on policy and strategy courses more than acquisition and data analysis. However, some courses do incorporate methods and tools, with some systems analysis. NDU also offers a course on big data, designed in response to the FY 2017 NDAA.

We categorized courses related to data and analysis at DoD Institutions

- Overall, the defense universities teach many acquisition-applied data analytics courses along with their concentrations in acquisition theory, processes, methods, and tools
  - More than 50 percent of the courses for NPS contain acquisition-applied data analytics
  - DAU concentrates on acquisition theory (including processes, methods and tools), with some emphasis on acquisition-applied data analytics
  - AFIT School of Engineering & Management (E&M) offers programs that teach data science and have relevance to acquisition-applied data analytics
  - AFIT School of Systems & Logistics (S&L) offers acquisition courses with less emphasis on data analytics
  - NDU has mostly courses on strategy that require little or no quantitative data

	Acquisition Theory and Processes Only	Acquisition Theory, Processes and Methods	Acquisition Processes & Methods w/ Tools	Acquisition-Applied Data Analytics	General Data Analytics	Other (not acquisition or data related)	Total Number of Courses
DAU	56	68	41	49	9	0	223
NDU	0	12	15	9	6	109	151
NPS	3	12	5	31	6	2	59
AFIT (S&L)	38	2	4	23	3	25	95
AFIT (E&M)	2	9	5	27	1	6	50
TOTAL	**99**	**103**	**70**	**139**	**25**	**142**	**578**

Increasing analytical acquisition courses

RAND

Approach

To determine what sorts of data analytics courses students at the government universities are exposed to, we first developed a set of categories by which to classify each course. The categories are designed to provide some insight into the level of data analytics the students would receive from an individual course.

The first category is acquisition theory and processes only. As the name suggests, students learn about acquisition theory and processes in these courses. These courses are foundational for students entering the AWF. A course example from this category is DAU's ACQ 340, Advanced International Management Workshop. The second category also includes methods. In these courses, students learn how to perform acquisition tasks. A course example in this category is AFIT's LOGM 619, Transportation Policy and Strategic Mobility. In the third category of classes, students are introduced to tools that are used in acquisition tasks. For example, DAU has dozens of tools, all of which are available on the DAU website, for use in acquisition. A course example in this category is DAU's BCF 209, Acquisition Reporting Course, Part A.

In the fourth category are courses that expose students to applied data analytics for acquisition. These courses involve such activities as determining the problem, collecting data, using tools or building models, and interpreting outputs. An example of a course in this category is DAU's BCF 330, Advanced Concepts in Cost Analysis. Courses that fall into a general data analytics category were counted separately. NPS's CS3315, Introduction to Machine Learning and Big Data, is an example of a course in this category. More-fundamental courses, such as Statistics, in which students learn to analyze data but without a specific acquisition focus, are also counted in this category.

Finally, we counted the number of courses that are unrelated to acquisition or data that appear in the course catalogs (listed in the "other" column in the previous slide). Some courses do not fit the acquisition topic or are not data related. An example of this is NDU's CISA 6916, Rule of Law: Policing and National Security.

Overall Findings

As can be seen in the previous slide, the institutions offer different types of analytic courses. This is to be expected, as each institution has its own primary functions. DAU, which offers the largest number of acquisition courses, at 223, includes courses of all types (except the "other" category), as has been defined for this study: 124 of the 223 courses we analyzed teach theory, processes, and some methods. The remaining 99 courses explore acquisition tools, applied analytics, and some general data analytics more deeply.

NPS offers many courses that fit the applied analytics category, accounting for more than one-half of the NPS courses analyzed (NPS, undated-a, undated-b). NPS has also begun offering courses on big data (NPS, 2018, undated-a).

AFIT also has a number of applied and general data analytics courses. At AFIT, 54 of the 145 courses analyzed were put into data analytics categories of general purpose or acquisition.

NDU primarily offers courses that have little emphasis on quantitative data analysis (which fits its institutional purpose). However, there are courses, 42 of the 151 analyzed, that teach some quantitative methods: 15 of the 151 fit into the general or applied data analytics categories. Of the 578 courses analyzed, only 142 are not acquisition or data related.

Across DAU, NDU, NPS, and AFIT, nearly 75 percent of the courses analyzed are related to acquisition. About 28 percent of the courses analyzed were put into data analytics categories of general purpose or acquisition.

Examples of applied and general-purpose courses at DAU

Examples Out of 223 DAU Courses Analyzed and Categorized

Category	Example Courses	Description
Applied		
Acquisition Theory and Processes Only	Export Controls (CMQ 216)	Provides an overview of export control regulations, with a specific emphasis on how these regulations pertain to U.S. Government personnel when delegating contract surveillance to foreign persons.
Acquisition Theory, Processes and Methods	Engineering Support to Technical Reviews (CME 203)	Provides DCMA Engineers with a firm understanding of their roles and responsibilities in executing a three-phase, six-step methodology for providing effective program support to acquisition program technical reviews, using DCMA guidelines.
Acquisition Processes & Methods w/ Tools	Fraud Awareness (AUD 1283)	Overview of the auditor's responsibility for the consideration of fraud in DCAA's audits and to heighten auditor awareness of the possibility of fraudulent activities.
Acquisition-Applied Data Analytics	Mission-Focused Services Acquisition (ACQ 265)	This course is designed to improve our tradecraft in the acquisition of services. It uses a multifunctional approach that provides acquisition team members with the tools and techniques necessary to analyze and apply performance-based principles when developing requirements documents and effective business strategies for contractor-provided services.
	Cost Risk Analysis (BCF 206)	Cost analysts taking this course are given an overview of how to model the cost/risk associated with a defense acquisition program. Topics covered include basic cost risk concepts, subjective probability assessment, goodness-of-fit testing, basic simulation concepts, and spreadsheet-based simulation.
General		
Data Analytics	Introduction to Probability and Statistics (CLE 035)	This continuous learning module provides participants with a basic introduction and understanding of probability and statistics, a crucial foundation for the Test and Evaluation (T&E) career field.

RAND

Findings for Government Schools

Findings for DAU

In the slide above, we provide examples of courses in each category out of the 223 courses we analyzed in the DAU course catalog (DAU, undated-b). This type of examination was performed for all the courses at DAU, NDU, NPS, and AFIT. Most courses fall within the applied data analytics category. Examples of courses are Export Controls (CMQ 216), Engineering Support to Technical Reviews (CME 203), Fraud Awareness (AUD 1283), Mission-Focused Services Acquisition (ACQ 265), and Cost Risk Analysis (BCF 206). The slide shows the categorization and how the different course types fit within the categories. Additionally, a few general data analytics courses are offered. An example of a general data analytics course is Introduction to Probability and Statistics (CLE 035).

About a quarter of DAU's courses provide data analytics tools

Examples of Applied Analytic Tools in DAU Courses

Tool Name	Application
Ishikawa	Root Cause Analysis
Microsoft Project, Oracle Primavera	Integrated Master Schedule (IMS) analysis
Acquisition Requirements Roadmap Tool Cost Estimating Module	Services Acquisition contract cost estimating
ARRT Risk Management Module	Risk Analysis
ARRT Performance Management Module	Enables alignment between requirements, standards & government surveillance of contract
Empower	Earned-Value Management (EVM) Data Analysis
SteelRay	IMS Metrics
@Risk	Schedule Risk Analysis
5 Why's	Problem solving and root cause analysis
Affinity Diagram, Deployment Flowchart, Force Field Analysis, Histogram, Nominal Group Technique (NGT), Pareto Charts, Scatter Diagram, Risk Management Cube	Problem Solving, Root Cause and Risk Analysis
Prioritization Matrix	Risk prioritization analysis
Dragonfly Simulation	Analysis of manufacturing, budget, test and evaluation, and sustainment cost/benefit and risk
Oracle Crystal Ball	Schedule and cost risk analysis using Monte Carlo simulation

RAND

DAU has developed dozens of tools for use in the acquisition community. DAU provided us with insight into its tools and how its tools are used; several such tools (relevant to data analytics) are shown on the slide above, along with their applications. Many of these tools are used as part of coursework at DAU: According to information provided by DAU, 49 percent of all DAU courses use specific tools, although the types of tools vary. For example, DAU's ACQ 202, Intermediate Systems Acquisition, Part A, uses Ishikawa for root-cause analysis in the curriculum. Each of the tools listed in the slide has an application. We examined the tools and assessed that about one-half are applicable to data analytics.

NDU data analytics courses focus on management training and leadership decision-making

- NDU courses relate to Data Analytics and analysis in general
  - CIC 6004: Big Data to Decisions (EIT/DAV)
    - *Developed in response to the FY2017 NDAA*
    - *Course includes relevant Big Data readings*
  - CIC 6037: Data Analytics for Decision Makers (DAV)
  - BCP (6606): White House, Congress, and the Budget
  - DMS (6414): Data Management Strategies and Technologies: A Managerial Perspective
  - DAL (6420): Data Analytics for Leaders
  - DMS (6414): Data Management Strategies and Technologies: A Managerial Perspective
  - FFR (6607): The Future of Federal Financial Information Sharing
- The emerging Data Analytics and Visualization (DAV) program examines the need for information superiority and decision making through data and big data concepts, and data analytics
- NDU publications illustrate application to informing decision-making
  - e.g., Kimminau (2015) and Lester et al. (2018)

NOTE: EIT = Emerging Information Technologies; DAV = Data Analytics and Visualization.

Findings for NDU

NDU has introduced big-data courses into its curriculum (NDU, 2018). Big data (data that have volume, variety, velocity, and veracity) are too complex for traditional data analysis applications. The methods used to investigate big data also differ. Where traditional data analysis asks a question and seeks data to help answer it, big data flips the process around, using large amounts of data to help identify issues, trends, and other useful information. Data visualization tools, such as Tableau, help analysts identify trends. A major benefit of Tableau and similar tools is relative ease of use and heavy focus on visualization. Visualization allows analysts and others to see trends or patterns in data. Visualizing patterns in data can sometimes be easier using such tools as Tableau than viewing the data in tables or other formats. Graphs often illustrate oddities or trends in data that other forms of viewing cannot. However, using these tools requires knowledge on developing data sets. Tableau does not include options to perform extract, transform, and load or extract, load, and transform. Data must be accessible in a "completed" format for Tableau to load and visually display (Freakalytics, 2014). Examples of NDU examinations of data analytics to support leadership decisionmaking are publications by Kimminau (2015) and Lester et al. (2018).

NPS is a traditional university with general data science courses plus applied training embedded in acquisition courses

- Data science courses are available to students as electives
- Some acquisition fields (e.g., cost analysis) include extensive data science courses
- Not reasonable to expect all acquisition students would become data science specialist
  - Same goes for other domains
    - *e.g., quality management, systems engineering, test and evaluation, contracting*
  - Acquisition management courses include overviews of analytically related disciplines
    - *e.g., systems analysis methods and functional area concerns...requirements analysis...test and evaluation*

General data science courses are available to students but not mandatory in general acquisition curricula

Findings for NPS

NPS is a traditional university with general data analytics courses and applied training embedded in acquisition courses (NPS, 2018). Data analytics courses (e.g., OS4106, Advanced Data Analysis) are available to students as electives. Overall, we found that the core acquisition curriculum did not include general data analytics courses, but data analytics courses could be taken as part of a student's elective coursework. Additionally, courses were offered in applied data analytics for acquisition, such as cost estimation (OS3701, Cost Estimation I: Methods and Techniques). Some acquisition fields (e.g., cost analysis) include extensive data analytics courses. However, it is not reasonable to expect all acquisition students to become data analytics specialists. The courses at NPS are geared toward students who need exposure to various subjects. The same goes for other domains, such as quality management, systems engineering, T&E, and contracting. All courses help expose students and give them a general background. The acquisition management courses include overviews of analytically related disciplines, such as systems analysis methods and functional area concerns, requirements analysis, and T&E.

Example: NPS has an extensive curriculum on data analytics and other evaluation-related methods

- Heavily analytic degrees and programs
  - Masters of Cost Estimating and Analysis
    - Operations Research for Cost Analysis
    - Probability and Statistics I
    - Probability and Statistics II
    - Advanced Concepts in Cost Estimating
    - Risk and Uncertainty Analysis
    - Decision Analysis
    - Applied Cost Analysis
    - Cost Economics

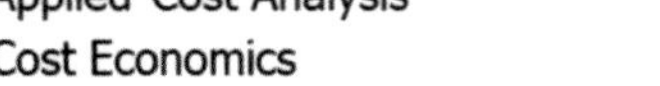

  - Data Science Certificate Program
    - Basic computation; Basic statistics and data analysis; Large data; Machine learning; Assessment; Supervised Learning; Unsupervised learning

- Analytic Courses (*partial list*)
  - OS4106 Advanced Data Analysis
  - OS4118 Statistical and Machine Learning
  - CC4920 Multi-Criteria Analysis
  - CS3315 Introduction to Machine Learning and Big Data
  - CS3636 Data Fusion with Online Information Systems
  - CS3640 Analysis of DoD Critical Infrastructure Protection
  - CS3695 Network Vulnerability Assessment and Risk Mitigation
  - CS3802 Computational Methods for Data Analytics
  - CS4312 Advanced Database Systems
  - CS4315 Introduction to Machine Learning and Data Mining
  - CS4678 Advanced Cyber Vulnerability Assessment
  - CS4680 Introduction to Risk Management Framework
  - CY3650 Cyber Data Management and Analytics
  - DA3450 Open Source Data Analysis
  - DA3610 Visual Analytics
  - DA4610 Dynamic Network Analysis
  - EC2100 Circuit Analysis
  - EC2110 Circuit Analysis II
  - EC2410 Analysis of Signals and Systems
  - EC3310 Optimal Estimation: Sensor and Data Association
  - EC3460 Introduction to Machine Learning for Signal Analytics
  - EC3500 Analysis of Random Signals
  - EC4747 Data Mining in Cyber Applications
  - GB1000 Quantitative Skills for Graduate Management Studies
  - GB3040 Managerial Statistics
  - GB3042 Process Analytics
  - GB3050 Financial Reporting and Analysis
  - GE3040 Statistics for Executive Management
  - GE3040 Statistics for Executive Management
  - GE3042 Process Analytics
  - GE3050 Financial Reporting and Analysis
  - IS3200 Enterprise Systems Analysis and Design
  - Etc.

http://nps.smartcatalogiq.com/en/Current/Academic-Catalog/Courses

NPS has an extensive curriculum on data analytics and other evaluation-related methods: The slide above shows a partial list of the individual courses. NPS offers two programs that are heavy in analytics. The first is the master in cost estimating and analysis. This degree has a number of courses in analytics, including probability and statistics, cost estimating, risk and uncertainty, and economics. The other program, the data science certificate, provides a range of data analytics courses, from basic statistics to machine learning.

AFIT is a traditional university that offers acquisition courses and data science courses

- AFIT School of Systems and Logistics offers a series of acquisition courses and workshops
  - Most are management-related
  - Relatively few have data science training
- AFIT School of Engineering and Management offers several acquisition-relevant programs, all of which include data science courses:
  - Cost Capability Analysis Certificate Program & M.S. Cost Analysis
  - M.S. Systems Engineering and M.E. Applied Systems Engineering
  - M.S. Logistics, M.S. Logistics and Supply Chain Management, & Supply Chain Management Certificate program
  - M.S. Operations

Findings for AFIT

AFIT is a traditional university that offers acquisition courses and data analytics courses. The School of Systems and Logistics offers a series of acquisition courses and workshops that are largely management related, with relatively few data analytics courses (AFIT, School of Systems and Logistics, undated). However, the School of Engineering and Management offers several acquisition-relevant programs, all of which include data analytics courses (AFIT, 2017). The School of Engineering and Management also offers several degrees and certificates: the cost capability analysis certificate and the supply-chain management certificate and the M.S. in cost analysis, M.S. in systems engineering, M.E. in applied systems engineering, M.S. in logistics, M.S. in logistics and supply-chain management, and M.S. in operations.

Example: AFIT School of Systems & Logistics Acquisition Courses Relevant to Data Analytics

Course Name	Course Type	Course Length
21X 312 Programming and Budgeting for Staff Logisticians	Instructor Led, Live Internet Course	15 days
21X 324 Deployment Planning	e-Learning Course	45 days
MRC 102 Mission Ready Contracting Officer Course	In-Residence or On-site Course	23 days
QMT 290 Integrated Cost Analysis	In-Residence or On-site Course	4 days
REL 210 Reliability Basics for Acquisition Professionals	In-Residence or On-site Course	
REL 310 Reliability and Reliability Growth Foundations I	In-Residence or On-site Course	5 days
REL 410 Reliability and Reliability Growth Foundations II	In-Residence or On-site Course	4 days
SOT 210 Science of Test Fundamentals for Managers	In-Residence or On-site Course	2 days
SOT 310 Science of Test: Experimental Design and Analysis I	In-Residence or On-site Course	4.5 days
SOT 410 Science of Test: Experimental Design and Analysis II	In-Residence or On-site Course	4.5 days
SYS 031 Intelligence in Acquisition Life Cycle Management	e-Learning Course	9 hours
SYS 110 Fundamentals of Data Management	e-Learning Course	10 hours
SYS 114 Acquisition Incident Review Process	e-Learning Course	45 days
SYS 150 Engineering Data Management	In-Residence or On-site Course	4 days
SYS 153 Early Tester Involvement	e-Learning Course	
SYS 182 Introduction to Systems Engineering	e-Learning Course	4 hours
SYS 208 Life Cycle Risk Management Course	In-Residence or On-site Course	3 days
SYS 213 Management of the Manufacturing Readiness Process	In-Residence or On-site Course	3 days
SYS 240 Avionics Cyber Vulnerability Assessment, Mitigation, and Protection	Instructor Led, Live Internet Course	2 days
SYS 252 Developmental Test & Evaluation (DT&E) High Performance Team Member Course	In-Residence or On-site Course	2 days/14 hours (Resident)
SYS 253 Early Test and Evaluation Influence in Acquisition	In-Residence or On-site Course	3 days
SYS 279 Logistics Assessment (LA) Assessor's Course	In-Residence or On-site Course	1.5 days
SYS 281 AF Acquisition and Sustainment Course	In-Residence or On-site Course	3 days
SYS 282 Management of the Systems Engineering Process	In-Residence or On-site Course	3 days
SYS 316 Advanced Air Worthiness Certification Course	In-Residence or On-site Course	8 days
SYS 383 Architecting in the Air Force	In-Residence or On-site Course	3 days
WKS 0657 Better Business Deals	In-Residence or On-site Course	3.5 days
WKS 0658 Data Analytics for the Rest of Us	In-Residence or On-site Course	2 days
WKS 0668 Applied Concepts of Data Analytics	In-Residence or On-site Course	24 hours
WKS 0669 Data Analytics for Senior Leaders	In-Residence or On-site Course	8 hours
WKS 0670 Professional Services Acquisition	Workshop	2 days
WKSP QMT 490 Current Topics in Cost Estimating	Workshop	2.5 days

RAND

The slide above shows a list of courses from AFIT's School of Systems and Logistics that are relevant to data analytics. Some, such as Fundamentals of Data Management, are offered online, but many need to be taken in person, which may present obstacles to professionals who wish to take these courses. However, with few exceptions, most courses last only a few days. AFIT also offers workshops, such as Current Topics in Cost Estimating.

DAU attendance shows large numbers of acquisition-applied data analytics training and some general data-analytic training

- **As expected, most of DAU's students take applied acquisition courses**
  - A significant number include tools (simpler calculations) as well as deeper acquisition-applied data analytics
- **Attendance in general data-analytic courses is in the thousands per year**

FY2017	Acquisition Theory and Processes Only	Acquisition Theory, Processes and Methods	Acquisition Processes & Methods w/ Tools	Acquisition-Applied Data Analytics	General Data Analytics	Other (not acquisition or data related)
AWF Civilian	23,611	48,188	35,320	40,200	3,125	0
AWF Military	4,145	6,972	6,555	7,790	104	0
Civilian	12,551	25,815	5,226	8,211	325	0
Military	6,534	12,351	1,805	4,213	120	0
Contractor	1,909	6,802	1,342	3,174	97	0
TOTAL	**48,750**	**100,128**	**50,248**	**63,588**	**3,771**	**0**

FY2018	Acquisition Theory and Processes Only	Acquisition Theory, Processes and Methods	Acquisition Processes & Methods w/ Tools	Acquisition-Applied Data Analytics	General Data Analytics	Other (not acquisition or data related)
AWF Civilian	22,253	49,927	36,640	37,381	2,640	0
AWF Military	4,038	7,661	7,007	7,444	142	0
Civilian	15,640	29,005	6,096	8,856	310	0
Military	9,032	13,203	1,784	3,813	141	0
Contractor	1,111	4,653	953	1,931	51	0
TOTAL	**52,074**	**104,449**	**52,480**	**59,425**	**3,284**	**0**

Increasing analytical acquisition courses →

CAUTION: There are about 10 times as many civilians in the AWF as military, so straight attendance comparisons are problematic. Also, we do not have attendance by individual to understand the extent of analytic training across the entire AWF.

RAND

DAU Course Attendance

In addition to obtaining the numbers of relevant courses at various educational institutions, we also obtained course attendance from DAU. The general attendance of DAU students, for FY 2017 and FY 2018, shows the distribution of students among course categories and the student type. Tracking is done by student type (military, civilian, or contractor) and by class type (online or classroom) and includes whether the student is from the AWF. The data show that the total number students enrolled in DAU increased between 2017 and 2018 but is roughly similar from year to year. At DAU, students tend to focus on courses that combine acquisition theory, processes, and methods. Many attendees enroll in DAU courses in acquisition-applied data analytics—approximately 64,000 and 59,000 in FY 2017 and FY 2018, respectively. An example DAU course in acquisition-applied data analytics is TST 303, Advanced Test and Evaluation. A comparatively small number of students also take general data analytics courses, such as CMQ 231, Data Collection and Analysis Application.

The tables in the slide above do not show data by individual student or class. Rather, the tables show the numbers of students who took courses that fall into each category. One individual could account for multiple counts in each class type category, so determining percentages is not appropriate. However, in FY 2018, more than 62,500 students took courses in the general and applied data analytics categories. It should be noted that the AWF has about ten times as many civilians as military personnel, so straight attendance comparisons are problematic. The data show that the ratio of AWF civilians to military in applied data analytics courses was about five to one, double the ten-to-one ratio that exists in the AWF workforce.

Through external institutions and established partnerships, DoD's workforce has access to additional data-analytics training

- Some specific examples of commercial institutions' programs in both data analytics and acquisition include:
  - *Georgia Tech* offers DAU-approved continuing education to government and acquisition professionals
    - Most courses are acquisition theory, process, methods, and tools related
  - *American University* offers the Key Executive Leadership Program in partnership with the DAU. Students in this program can attain an AU Master of Public Administration degree
    - Offers some courses involving data analysis, primarily analysis methods
  - *George Mason University* offers certificate programs for government contractors and acquisition professionals as well as an MPA in Public Administration. They also offer a MS, Data Analytics Engineering
    - Relatively few data analysis courses are present in the certificate programs; those present are primarily related to the cost discipline and are part of the Government Contracts Accounting Certificate
    - MPA Public Administration involves the highest number data analytics-specific courses relevant to acquisition out of all of the universities
  - *Georgetown University* offers a Master of Public Policy
    - Core classes are relevant to data science and various types of analysis
  - *George Washington University* offers an MS in Government Contracts
    - Has a few courses dealing with applied data analysis; however, most are acquisition theory and process related

Findings for Nongovernmental Institutions

We additionally examined several nongovernmental institutions to determine other educational opportunities relevant to data analytics and acquisition and found a variety of options accessible to the DoD workforce. DAU partners with universities to offer courses in acquisition. We found that Georgia Tech and American University, to name just two, offer relevant acquisition and data analytics coursework in partnership with DAU (Georgia Institute of Technology, undated; American University, undated). Several other universities in the Washington, D.C., area offer relevant programs. These programs range from shorter certificate programs, aimed at acquisition professionals, to master degree programs offering courses in various types of analysis. We also note that some universities offer programs in data analytics (e.g., GMU's MS in data analytics engineering); however, such programs are designed to be specific to data analytics and do not generally teach principles of acquisition (GMU, undated-a, undated-b, undated-c; Georgetown University, undated; George Washington University, undated). The list of partner universities is not exhaustive. We have provided some specific examples in this section to help characterize what is available.

External partnership institutions provide additional applied data analytics in acquisition

- GMU's MPA program has the highest number of courses in acquisition-applied data analytics
  - A few additional general data analytics courses are also available

	Acquisition Theory and Processes Only	Acquisition Theory, Processes and Methods	Acquisition Processes & Methods w/ Tools	Acquisition-Applied Data Analytics	General Data Analytics	Other (not acquisition or data related)	Total Number of Courses
GMU (CF&A)	5	0	0	1	0	0	6
GMU (CGA)	2	0	0	4	0	0	6
GMU (MPA)	0	1	0	8	2	7	18
Georgetown (MPP)	0	1	1	5	0	11	18
GA Tech	7	3	1	1	0	0	12
AU	2	2	0	4	0	7	15
GW (GC)	10	3	0	4	0	0	17
TOTAL	**26**	**10**	**2**	**27**	**2**	**25**	**92**

Increasing analytical acquisition courses

RAND

NOTE: GA Tech = Georgia Institute of Technology; AU = American University; GW = George Washington University; GC = master in government contracts.

As with the government institutions, we categorized the courses for each of the private-sector university programs with relevance to data analytics in acquisition. The programs include the following:

- GMU: certificate in contract formation and administration, certificate in government contracts accounting, MPA (GMU, undated-c)
- Georgetown University: MPP (Georgetown University, undated)
- Georgia Tech: Contracting Education Academy, offering continuing education in acquisition and government contracting (Georgia Institute of Technology, undated)
- American University: Key Executive Leadership Program—students in this program earn an MPA (American University, undated)
- George Washington University: master in government contracts (George Washington University, undated).

We collected data on available courses at each of these institutions, including the course name, course description, and relevant program, and categorized the courses as we did for the government school classes and as shown in the slide above. Examples of types of courses that fell into each category are listed as follows, along with the university that offers that course:

- Acquisition theory and processes only: Government Contracting Fundamentals (GMU)
- Acquisition theory, processes, and methods: Statistics for Managers (George Washington University)

- Acquisition processes and methods with tools: Fundamentals of the Federal Acquisition Regulation (Georgia Tech)
- Acquisition-applied data analytics: Managing in the Information Age (American University)
- General data analytics: Advanced Data Analysis for Policy and Government (GMU)
- Other (not acquisition or data related): Ethics, Values and Public Policy (Georgetown).

The overall total number of courses that appear relevant to acquisition is smaller than for DAU and other defense universities; however, these institutions offer opportunities in areas of interest to this study. GMU's MPA program and Georgetown's MPP program had the most courses that appeared to focus on acquisition or general data analytics. Most of the classes in Georgetown's program were not relevant to data or acquisition: Of the seven courses that were relevant, five were in the acquisition-applied data analytics category. Conversely, 11 of the 18 courses in GMU's MPA program were relevant to data analytics or acquisition, and a majority of courses offered in George Washington University's MS in government contracts were relevant to acquisition theory or data analytics. American University's program was almost evenly split between acquisition- or data-relevant courses and those not related to either acquisition or data.

Regarding the certificate and continuing education programs offered by GMU and Georgia Tech, we found that all courses were relevant to data or acquisition. Georgia Tech's program is mostly theory, with some courses that combine theory, processes, and methods; one course that involves the use of data analytics tools; and one that involves applied data analytics. GMU's certificate in contract formation and administration primarily comprises courses in acquisition theory, whereas GMU's certificate in government contracts accounting primarily comprises courses in applied data analytics, with two courses in acquisition theory.

DAU Partnerships with Other Institutions

DAU also leverages strategic educational partnerships to provide opportunities to the workforce that DAU cannot support in-house

Course	Provider
Data Scientists Toolbox	Johns Hopkins University
Developing Data Products	Johns Hopkins University
R Programming	Johns Hopkins University
Machine Learning Foundations	Stanford
Applied Text Mining in Python	University of Michigan
Serverless Data Analysis with Google BigQuery	Google
Data Bases and SQL for Data Science	IBM
Cyber Training	DC3 Cyber Training Academy

Most are general data analytics courses

RAND

NOTE: SQL = Structured Query Language.

In addition to in-house courses offered by DAU, the university has identified a number of outside organizations from which to leverage higher-level expertise in data analytics (DAU, personal communication). Several years ago, DAU faculty observed the increasingly prominent role of data analytics in the acquisition process and noted the need to expand. The table in the slide above shows courses that DAU offers via partnerships with Johns Hopkins University, Stanford University, University of Michigan, Google, IBM, and DC3 Cyber Training Academy. Johns Hopkins University is being tasked with teaching DAU students R programming, data scientist toolboxes, and how to develop data products. DAU has also been able to leverage similar classes taught by skilled Johns Hopkins faculty. Similarly, Stanford is being utilized to teach machine learning foundations, University of Michigan is offering text mining using Python, Google offers its BigQuery tool, IBM offers classes in SQL databases, and DC3 Cyber Training Academy provides cyber training. Although attendance is not tracked, these courses offer deeper understanding of data analytics tools, modeling methods, databases, and other important areas of knowledge that track back to the data governance processes shown earlier, which were lacking at DAU. Again, these examples illustrate relevant courses and institutions but do not constitute an exhaustive list of all opportunities.

Thus, the Department of Defense acquisition workforce has access to data analytics training through multiple avenues

- DoD's acquisition training institutions have expanded their data analytics offerings
  - Data governance, database operations, data management, and data security are largely present in the curriculums of the universities
  - Some areas (e.g., data collection, data development, data transformation, data quality management, data architecture management) appear to be lacking in these institutions
- Understanding that DoD should be moving toward the use of advanced analytics in decision-making, DoD's institutions have taken multiple steps to improve offerings, including:
  - Additional partnerships with organizations at the forefront of advanced analytics (e.g., Google)
  - In 2018, NPS stood up a new interdisciplinary working group to streamline the way data science is developed, taught, and shared
- However, we did not assess the quality of the curriculum

Summary of Data Analytics Training Opportunities and Challenges

We found multiple avenues of data analytics training to be available to the AWF, through government universities and private-sector institutions. Recognizing the advantages of applying data analytics to acquisition, institutions are working to expand offerings in both fields. DAU has partnered with such institutions as IBM and Google to offer data courses (DAU, personal communication). The courses are available to students and expand the data analytics offerings from DAU. Additionally, many private-sector universities and other commercial institutions offer DAU-approved courses; students taking these can earn DAU credit (DAU, undated-a). NDU has also expanded offerings to include a course on big data; the big-data field offers advancements in data analytics capabilities to the DoD. Furthermore, NPS stood up an interdisciplinary data analytics working group, the Data Science and Analytics Group, in 2018. The new working group aims to organize existing data sources and provide insights on how to teach data analytics, how to develop human capital for the data analytics field, and how to better use data analytics for decisionmaking (Schehl, 2018).

An informed and well-trained workforce might be one of the greatest challenges that the acquisition community faces. DoD analysts need training and licenses for analytic tools beyond Excel. The AWF also needs training to become consumers of broader analysis. Providing the workforce tools and data is useful only if the workforce can effectively use the tools and data to help inform decisions.

This analysis has several limitations. One is that we did not assess whether these courses are sufficient. There appear to be many opportunities covering a wide range of subjects, but it is difficult to judge whether this offering is the correct number of courses, or if even more courses or additional subjects are needed to provide sufficient training. Through discussions with DoD subject-matter experts, we found that the DoD's military and civilian employees might not be fully aware of the opportunities available to them through these universities. Likewise, it is not clear whether the workforce has the time and support to attend these courses. Not all the courses are online, and some would require a time commitment away from employees' day jobs. To help alleviate the need for that time away, DAU has moved toward offering additional online courses to improve access. Another limitation is that we did not research every possible defense university. Army subject-matter experts mentioned that the U.S. Army War College is training Army leadership (e.g., colonels), who are then in the field determining requirements for future weapon systems. An additional conclusion based on our discussions is that training is likely even more widespread than noted here and largely decentralized.

4. Conclusions

What's in this Briefing

- Background
- Approach and methodology
- Findings on congressional questions
- Conclusions

Potential Opportunities by Stakeholder

Finally, we grouped our ideas of potential opportunities and actions identified in our study according to what stakeholder group would implement them.

Congress

Two opportunities generally apply for congressional consideration. First, Congress could clarify in 10 U.S.C. 2222(e) that all acquisition and sustainment data are *common enterprise data* and thus available across the DoD. Doing so would provide additional statutory authority for access and may improve sharing. Second, Congress could make permanent the three-year pilot established by Section 235 of the FY 2017 NDAA, which grants analysts in FFRDCs access to sensitive data.[15] This pilot provides an efficient mechanism to establish access to, and protection of, sensitive data beyond government employees.

DoD Leadership

DoD leadership (USD[A&S], with the CMO, the CIO, CDOs, and SAEs) could consider several options:

- First, leadership could address disincentives for data sharing, including clarifying oversight extent and roles to minimize concerns of micromanagement above the PEO,

[15] For full disclosure, the authors work for an FFRDC, so the quality of our research would benefit from such access. Also note that the access problems extend to nongovernment analysts beyond FFRDCs.

expanding security metalabeling at the data-element level (rather than for a whole corpus of data), and further addressing security concerns by clarifying sensitivity upgrades from data aggregation.

- Second, leadership could direct full access to acquisition data across all DoD systems for analysts who are assessing broader institutional trends and performance correlates; this need is especially salient for lower-sensitivity data (e.g., predecisional information).
- Third, analytic tool access could be facilitated through wider use of virtual computing environments for analysis and an embellished list of standard software approved for installation on DoD computers.
- Fourth, R&D could continue to improve data and analytic systems and to develop new applications.
- Finally, strategy could be expanded beyond individual functional domains to develop cross-domain data and analysis strategy to integrate these kinds of improvements.

DoD Information Managers

For the DoD's information managers, the opportunities appear to surround continuation of the efforts to date and expansion to all core acquisition functional areas. These include (with leadership support and funding) continued pursuit of a PM and process suite that could automatically collect and distribute relevant data; continued maturation of data collection, access, and analytic layers on data systems; and continued compiling and sharing of information about where data reside and how to access them.

Defense Acquisition Training Institutions

Similarly, the defense acquisition training institutions could continue offering up-to-date applied and general data analytics courses, both as part of the core training curriculum and as electives to improve the DoD's organic AWF capabilities.

Summary Conclusions

Data analytics in the DoD support a broad range of acquisition functions. The DoD is using a mixture of advanced data analytics, COTS tools, and simpler approaches. What matters is using appropriate techniques to inform acquisition decisions, not whether a particular technique is more advanced or has been used for a long time. Whether the mixture employed by DoD is optimal is beyond the scope of this study, but we recognize that the DoD has many useful applications and is continuing to explore new techniques to ascertain their appropriateness, utility, and cost.

Data analytics are used to inform DoD decisionmaking throughout the acquisition community and across acquisition functions. These analyses inform acquisition management, insight, oversight, execution, and decisions at all levels, including program managers, contracting officers, engineers, auditors, oversight executives, and DoD leadership. However, data analysis may or may not be equally weighted against other considerations by decisionmakers. At times, other priorities (e.g., politics, optics) will override analysis. Evidence shows that data analytics results are sometimes not directly followed by decisionmakers for other reasons, and some of these decisions ended up being problematic in hindsight. Two possible reasons among many that decisionmakers choose another course of action are that the analysis was wrong or that it was not persuasive. This point illustrates the difficulty in trying to tie analysis to decisions, but we were able to review some well-documented problematic programs to see whether their Nunn-McCurdy root-cause analyses indicated where information was available but was not fully acted on by leadership. According to the Office of Performance

Assessments and Root Cause Analyses, such cases are limited to about one-quarter of problematic programs (USD[AT&L], 2016a) and involve mitigating incentives, strategic risk taking, and natural limitations when human judgment is involved.

The DoD has made progress in improving its data analysis capabilities and access to enabling data, including implementations of the latest commercial analytic tools, on top of open information systems. The DoD is also pursuing commercial best practices in data collection and management as it evolves its information systems in its main lines of acquisition functions. However, more needs to be done. The DoD is continuing to pursue investments and improvements along many of these best practices.

Although some improvement has been seen in data sharing, barriers are still problematic. Issues include a culture of data restriction, security concerns, lack of trust that data will be used appropriately (e.g., historical tendencies to staff invasiveness, micromanagement, and what economists call "moral hazards" [actions with side consequences not borne by the actor]), and the burden of data reporting. Also, hard questions still require analysis specialists who understand the acquisition domain in question and can involve extensive data collection, cleaning, and management. As with the private sector, only a limited pool of experts understands both acquisition and data analytics. In addition, recent breakthroughs in advanced analytics make it tempting to think they can easily be applied broadly to the DoD's problems, but they might not apply for various reasons. For example, we do not expect that clustering and machine learning tools can be directly applied to predict program performance: Programs fail for different reasons, making such failure something different from a clustering problem. Nevertheless, research is being funded at universities and laboratories in a search for useful applications.

Finally, we must scope our expectations of what data analytics can do for acquisition outcomes. The (albeit spectacular) recent commercial breakthroughs in data analytics do not always apply to the DoD, with its different mission and complicated incentives and constraints. Also, we must recognize that data analytics span a broad range. Easier efforts can involve relatively simple analysis and visualization of readily available data (e.g., business intelligence data). Deeper, more-complicated insights require other data over long periods that must be cultivated, managed, cleaned, and archived, requiring extensive and ongoing work. They cannot be constructed in a few weeks but rather require strategic planning and investing. We must also remember that data analysis is important for informing management and decisionmaking, but it cannot eliminate the risks inherent in acquisition—especially developmental defense acquisition that is rushing to deliver beyond-state-of-the-art capabilities to warfighters to gain a (temporary) competitive technical advantage. Data analysis is no magic bullet, but it may offer a range of bullets that are quite useful and can become more useful. The DoD is on that path but needs continued support and rational expectations. Finally, an implicit assumption also exists that better analysis means better outcomes. An important goal is to have better-informed decisionmakers who will make better decisions that result in better outcomes. This distinction emphasizes the particular importance of the workforce part of this challenge.

Steps for Future Research

Steps for Future Research

- **For Congress**
  - Identify DoD acquisition leadership structures that streamline acquisition while balancing decision incentives and authorities
  - Identify how statutes can be changed to give efficient access to sensitive data for UARCs, contractor labs, and SETA support contractor analysts while providing appropriate data protections
- **USD(A&S) with CMO, CIO, CDOs, SAEs**
  - Identify how to address disincentives to data sharing
  - Develop policies and processes for analysis on data aggregation and classification upgrades
  - Develop policies and approaches for granting DoD-wide access to different DoD information systems for government and contractor analysts
  - Identify the minimum data needed at what level and for what purposes given costs and benefits
  - Conduct detailed analysis to create a cross-domain DoD data-analytics strategy
- **For DoD acquisition training institutions**
  - Conduct an assessment of the quality and practical utility of data-analytics courses
- **New or expanded analyses**
  - Explore better ways to objectively separate effects of uncertainties and externalities in sustainment metrics
  - Recognize and explore mission-level (versus program-level) analysis
  - Examine better ways to utilize framing assumptions and their metrics
  - Conduct institutional performance analysis
  - Identify core data needed to answer important questions

Many of the options identified are concepts that require further research to develop specific implementation approaches.

Although the ideas outlined on the slide above provide some specificity into potential opportunities for improving data analytics in the DoD, some of these and other ideas require further research to develop actionable recommendations for implementation.

Congress

For Congress, sometimes other considerations led to acquisition decisions that ran counter to particular available data analytics results, apparently because of conflicting incentives and other strategic motivations. Research could further clarify what drives these decisions and how Congress could better balance decision incentives and authorities. For example, what is the right balance between streamlined oversight and independent checks and balances? Also, a broader need exists beyond FFRDCs for efficient access to (and protection mechanisms for) other sensitive data for analysis. Research is needed to develop options for UARCs, contractors working for defense laboratories, and SETA support contractors.

DoD Leadership

Further research could also inform actions by DoD leadership:

- how to address disincentives to data sharing

- policy and process analysis on data aggregation and classification upgrades, including aggregation of sensitive, unclassified information to ensure more-consistent application across the DoD
- policies and approach for granting DoD-wide access to different DoD information systems for government and contractor analysts
- identifying the minimum data needed at what level and for what purposes
- detailed analysis to create a cross-domain DoD data-analytics strategy.

Acquisition Training Institutions

For acquisition training institutions, the DoD needs to understand not only what types of data analytics courses are offered but also their quality and practical utility, especially given the diverse nature of the AWF and its range of technical backgrounds and skills.

Data Analysis Topics

Finally, expanding the type of data analysis topics has the potential to improve acquisition outcomes. Although the topics depend on current leadership priorities and needs, the following examples illustrate some possible areas that could be further explored.

- **Externalities:** Some data analytics results do not delineate the effects of external factors from those internal to the acquisition system. For example, operation and sustainment cost growth can result from poor fuel efficiency (a technical and design factor within the acquisition community's area of responsibility) or from increases in fuel cost (external to the DoD, from the global marketplace). Improved data analytics capabilities and approaches could be developed to try to separate these factors to inform decisionmakers on what is driving cost growth and whether the acquisition community is responsible or not.
- **Mission-level (versus program-level) portfolio analysis:** Outcome-based portfolio analysis could be performed (e.g., to understand technical, budgetary, and schedule implications to mission or kill-chain effects from cross-program dependencies). The DoD is exploring how to shift from assessing acquisition program performance in isolation to assessing performance as it pertains to the integrated set of programs that field mission-level capabilities (e.g., kill chains). Such analysis offers the potential to view acquisition from the perspective of integrated warfighter capability effects and outcomes rather than delivery of individual systems. Integrated analysis could help decisionmakers better understand dependencies among systems so that technical, budgetary, and schedule decisions maximize the integrated warfighting capability. For example, such a shift could enable better budgetary, schedule, and quantity decisions based on how a smaller program fits within an overall kill chain. Otherwise, smaller programs may receive insufficient support because leadership and Congress do not have the full picture of the importance of that program in larger warfighter capabilities.
- **Framing assumptions:** Framing assumptions, analysis, and associated metrics tracking could be fully implemented (Arena et al., 2013; Arena and Mayer, 2014; DoDI 5000.02, 2017). The DoD and its FFRDCs have developed how to assess the key conceptual risks of a program (Arena et al., 2013; Arena and Mayer, 2014) so that leadership can

understand the risk (i.e., the "big bets") they are taking when approving a program. Although codified in current DoD policy (DoDI 5000.02, 2017), this method could be expanded to allow Congress to better understand the risks involved in a new program it authorizes and funds. Also, it is unclear whether the current trends away from such approaches (e.g., in the "Middle Tier" of acquisition [10 U.S.C. 2302]) need to be clarified so that the utility of understanding the framing assumptions in a program at initiation is assessed and articulated to decisionmakers. Finally, tracking these risks could be made more explicit so that early indicators of fundamental conceptual risks are detected as soon as possible.

- **Performance analysis:** The DoD could continue making inroads into using data analysis to understand institutional performance trends and what may be driving them. Such quantitative and qualitative analyses seek to understand statistically significant trends rather than focusing on the performance of an individual case or program. For example, the DoD published four reports of such analyses since 2013 (see USD[AT&L], 2013, 2014, 2015, 2016a) and is rebuilding a capability to conduct such analysis in the OUSD(A&S). A particular challenge worth highlighting is the general lack of (and need for) more-predictive (leading) indicators of acquisition performance problems. Most available indicators used by the DoD are descriptive (lagging), although there are some examples of leading indicators, such as contractor risk indicators, earned value, and cross-program performance indicators (e.g., extrapolating known concerns to similar acquisition activities). There are no easy candidates for new leading indicators, so this is a research challenge, not an immediate opportunity.
- **Data needed:** Data collection and management involve significant effort and investments, so analysis to better understand which data are needed can inform acquisition efficiency efforts and ensure that key data to inform decisions and data analytics are available. Such efforts start with the important questions and decisions that need to be informed, but analysis includes assessing the relative costs and benefits of obtaining data through alternative mechanisms.

References

AFIT—*See* Air Force Institute of Technology.

Air Force Institute of Technology, *Graduate School of Engineering and Management 2017–2018 Catalog*, Wright-Patterson Air Force Base, Ohio, 2017. As of November 12, 2018: https://www.afit.edu/docs/2017-2018%20AFIT%20Graduate%20Catalog.pdf

Air Force Institute of Technology, School of Systems and Logistics, "Course Information and Registration," webpage, undated. As of November 12, 2018: https://www.afit.edu/LS/courselist.cfm?courseType=ACQ

Amazon Web Services, "What Is a Data Lake?" webpage, undated. As of May 3, 2019: https://aws.amazon.com/big-data/datalakes-and-analytics/what-is-a-data-lake/

American University, "Key Executive Leadership MPA Courses," webpage, undated. As of November 12, 2018: https://www.american.edu/spa/key/courses.cfm

Anton, Philip S., Tim Conley, Irv Blickstein, Austin Lewis, William Shelton, and Sarah Harting, *Baselining Defense Acquisition*, Santa Monica, Calif.: RAND Corporation, RR-2814-OSD, forthcoming.

Arena, Mark V., Irv Blickstein, Abby Doll, Jeffrey A. Drezner, Jennifer Kavanagh, Daniel F. McCaffrey, Megan McKernan, Charles Nemfakos, Jerry M. Sollinger, Daniel Tremblay, and Carolyn Wong, *Management Perspectives Pertaining to Root Cause Analyses of Nunn-McCurdy Breaches*, Vol. 4: *Program Manager Tenure, Oversight of Acquisition, Category II Programs, and Framing Assumptions*, Santa Monica, Calif.: RAND Corporation, MG-1171/4-OSD, 2013. As of February 27, 2019: https://www.rand.org/pubs/monographs/MG1171z4.html

Arena, Mark V., and Lauren A. Mayer, *Identifying Acquisition Framing Assumptions Through Structured Deliberation*, Santa Monica, Calif.: RAND Corporation, TL-153-OSD, 2014. As of February 27, 2019: https://www.rand.org/pubs/tools/TL153.html

Aronin, Benjamin S., John W. Bailey, Ji S. Byun, Gregory A. Davis, Cara L. Wolfe, Thomas P. Frazier, and Patricia F. Bronson, *Expeditionary Combat Support System: Root Cause Analysis*, Alexandria, Va.: Institute for Defense Analyses, October 2011. As of March 9, 2019: https://apps.dtic.mil/docs/citations/ADA571034

ASA(FMC)—*See* Assistant Secretary of the Army for Financial Management and Comptroller.

ASAF(FMC)—*See* Assistant Secretary of the Air Force for Financial Management and Comptroller.

ASN(FMC)—*See* Assistant Secretary of the Navy for Financial Management and Comptroller.

Assistant Secretary of the Air Force for Financial Management and Comptroller, *Department of Defense Fiscal Year (FY) 2019 Budget Estimates: Air Force, Justification Book of Research, Development, Test & Evaluation, Air Force RDT&E*, Vol. I, Washington, D.C.: U.S. Department of Defense, February 2018a.

———, *Department of Defense Fiscal Year (FY) 2019 Budget Estimates: Air Force, Justification Book of Research, Development, Test & Evaluation, Air Force RDT&*E, Vol. II, Washington, D.C.: U.S. Department of Defense, February 2018b.

———, *Department of Defense Fiscal Year (FY) 2019 Budget Estimates: Air Force, Justification Book of Research, Development, Test & Evaluation, Air Force RDT&E*, Vol. IIIa, Washington, D.C.: U.S. Department of Defense, February 2018c.

———, *Department of Defense Fiscal Year (FY) 2019 Budget Estimates: Air Force, Justification Book of Research, Development, Test & Evaluation, Air Force RDT&E*, Volume IIIb, Washington, D.C.: U.S. Department of Defense, February 2018d.

Assistant Secretary of the Army for Financial Management and Comptroller, *Department of Defense Fiscal Year (FY) 2019 Budget Estimates: Army, Justification Book of Research, Development, Test & Evaluation, Army RDT&E*, Vol. I: *Budget Activity 1*, Washington, D.C.: U.S. Department of Defense, February 13, 2018a.

———, *Department of Defense Fiscal Year (FY) 2019 Budget Estimates: Army, Justification Book of Research, Development, Test & Evaluation, Army RDT&E*, Vol. II: *Budget Activity 2*, Washington, D.C.: U.S. Department of Defense, February 13, 2018b.

———, *Department of Defense Fiscal Year (FY) 2019 Budget Estimates: Army, Justification Book of Research, Development, Test & Evaluation, Army RDT&E*, Vol. III: *Budget Activity 3*, Washington, D.C.: U.S. Department of Defense, February 13, 2018c.

———, *Department of Defense Fiscal Year (FY) 2019 Budget Estimates, Army, Justification Book of Research, Development, Test & Evaluation, Army RDT&E*, Vol. IV: *Budget Activity 4*, Washington, D.C.: U.S. Department of Defense, February 13, 2018d.

———, *Department of Defense Fiscal Year (FY) 2019 Budget Estimates: Army, Justification Book of Research, Development, Test & Evaluation, Army RDT&E*, Vol. V: *Budget Activity 5*, Washington, D.C.: U.S. Department of Defense, February 13, 2018e.

———, *Department of Defense Fiscal Year (FY) 2019 Budget Estimates: Army, Justification Book of Research, Development, Test & Evaluation, Army RDT&E*, Vol. VI, *Budget Activity 6*, Washington, D.C.: U.S. Department of Defense, February 13, 2018f.

———, *Department of Defense Fiscal Year (FY) 2019 Budget Estimates, Army, Justification Book of Research, Development, Test & Evaluation, Army RDT&E*, Vol. VII: *Budget Activity 7*, Washington, D.C.: U.S. Department of Defense, February 13, 2018g.

Assistant Secretary of the Navy for Financial Management and Comptroller, *Department of Defense Fiscal Year (FY) 2019 Budget Estimates: Navy, Justification Book of Research, Development, Test & Evaluation, Navy RDT&E*, Vol. I: *Budget Activities 1, 2, and 3*, Washington, D.C.: U.S. Department of Defense, February 13, 2018a.

———, *Department of Defense Fiscal Year (FY) 2019 Budget Estimates: Navy, Justification Book of Research, Development, Test & Evaluation, Navy RDT&E*, Vol. II: *Budget Activity 4*, Washington, D.C.: U.S. Department of Defense, February 2018b.

———, *Department of Defense Fiscal Year (FY) 2019 Budget Estimates: Navy, Justification Book of Research, Development, Test & Evaluation, Navy RDT&E*, Vol. III: *Budget Activity 5*, Washington, D.C.: U.S. Department of Defense, February 2018c.

———, *Department of Defense Fiscal Year (FY) 2019 Budget Estimates: Navy, Justification Book of Research, Development, Test & Evaluation, Navy RDT&E*, Vol. IV: *Budget Activity 6*, Washington, D.C.: U.S. Department of Defense, February 2018d.

———, *Department of Defense Fiscal Year (FY) 2019 Budget Estimates: Navy, Justification Book of Research, Development, Test & Evaluation, Navy RDT&E*, Vol. V: *Budget Activity 7*, Washington, D.C.: U.S. Department of Defense, February 2018e.

Bertolucci, Jeff, "Big Data Analytics: Descriptive vs. Predictive vs. Prescriptive," *Information Week*, December 31, 2013.

Blickstein, Irv, Michael Boito, Jeffrey A. Drezner, James Dryden, Kenneth Horn, James G. Kallimani, Martin C. Libicki, Megan McKernan, Roger C. Molander, Charles Nemfakos, Chad J. R. Ohlandt, Caroline Reilly, Rena Rudavsky, Jerry M. Sollinger, Katharine Watkins Webb, and Carolyn Wong, *Root Cause Analyses of Nunn-McCurdy Breaches*, Vol. 1: Zumwalt-*Class Destroyer, Joint Strike Fighter, Longbow Apache, and Wideband Global Satellite*, Santa Monica, Calif.: RAND Corporation, MG-1171/1-OSD, 2011. As of May 21, 2019:
https://www.rand.org/pubs/monographs/MG1171z1.html

Blickstein, Irv, Jeffrey A. Drezner, Martin C. Libicki, Brian McInnis, Megan McKernan, Charles Nemfakos, Jerry M. Sollinger, and Carolyn Wong, *Root Cause Analyses of Nunn-McCurdy Breaches*, Vol. 2: *Excalibur Artillery Projectile and the Navy Enterprise Resource Planning Program, with an Approach to Analyzing Program Complexity and Risk*, Santa Monica, Calif.: RAND Corporation, MG-1171/2-OSD, 2012. As of May 21, 2019: https://www.rand.org/pubs/monographs/MG1171z2.html

BLS—*See* U.S. Bureau of Labor Statistics.

Bolten, Joseph G., Robert S. Leonard, Mark V. Arena, Obaid Younossi, and Jerry M. Sollinger, *Sources of Weapon System Cost Growth: Analysis of 35 Major Defense Acquisition Programs*, Santa Monica, Calif.: RAND Corporation, MG-670-AF, 2008. As of October 17, 2018: https://www.rand.org/pubs/monographs/MG670.html

Caralli, Richard, Mark Knight, and Austin Montgomery, *Maturity Models 101: A Primer for Applying Maturity Models to Smart Grid Security, Resilience, and Interoperability*, Pittsburgh, Pa.: Carnegie Mellon University, November 2012. As of February 26, 2018: https://resources.sei.cmu.edu/asset_files/WhitePaper/2012_019_001_58920.pdf

DalleMule, Leandro, and Thomas H. Davenport, "What's Your Data Strategy?" *Harvard Business Review*, May–June 2017, pp. 112–121.

Data Cabinet, *The Federal Government Data Maturity Model*, Washington, D.C.: National Technical Information Service, undated. As of May 21, 2019: https://www.ntis.gov/assets/FDMM.pdf

DAU—*See* Defense Acquisition University.

Defense Acquisition University, "Defense Acquisition University Equivalency Program," webpage, undated-a. As of February 13, 2019: http://icatalog.dau.mil/appg.aspx

———, "Online Catalog," webpage, undated-b. As of November 12, 2018: http://icatalog.dau.mil/onlinecatalog/tabnav.aspx?tab=ACQ

———, *DAU Program Managers Tool Kit*, 12th ed., Fort Belvoir, Va., December 2002.

———, *Joint Program Management Handbook*, Fort Belvoir, Va.: Defense Acquisition University Press, October 2004. As of December 19, 2017: http://www.dtic.mil/docs/citations/ADA437767

———, *DAU Glossary*, Fort Belvoir, Va., February 9, 2017a. As of March 18, 2018: https://www.dau.mil/tools/t/DAU-Glossary

———, *Defense Acquisition Guidebook*, Fort Belvoir, Va., 2017b.

Defense Technical Information Center, DoD Investment Budget Search (R-2s and P-40s), database, undated.

Department of Defense Instruction 5000.02, *Operation of the Defense Acquisition System*, incorporating Change 3, Washington, D.C.: U.S. Department of Defense, January 7, 2015, effective August 10, 2017. As of December 17, 2017: http://www.esd.whs.mil/Portals/54/Documents/DD/issuances/dodi/500002_dodi_2015.pdf?ver=2017-08-11-170656-430

DiCicco, Ralph, "AVWG Update," briefing, U.S. Air Force, Chief Acquisition Capability Division Department, Washington, D.C., February 19, 2019, Not available to the general public.

Director of Performance Assessments and Root Cause Analyses, "PARCA Root Cause Analysis for the DDG-1000 Program," Washington, D.C.: Office of the Assistant Secretary of Defense for Acquisition, memorandum for Under Secretary of Defense for Acquisition, Technology, and Logistics, May 21, 2010a.

———, "PARCA Root Cause Analysis for the Joint Strike Fighter (JSF) Program," Washington, D.C.: Office of the Assistant Secretary of Defense for Acquisition, memorandum for Under Secretary of Defense for Acquisition, Technology, and Logistics, May 25, 2010b. As of February 23, 2019: https://www.acq.osd.mil/aap/assets/docs/2010-05-25-parca-rca-jsf.pdf

———, "PARCA Root Cause Analysis for the Enterprise Resource Planning Systems," Washington, D.C.: Office of the Assistant Secretary of Defense for Acquisition, memorandum for Under Secretary of Defense for Acquisition, Technology, and Logistics, February 23, 2011a.

———, "PARCA Root Cause Analysis of the Global Hawk Program," Washington, D.C.: Office of the Assistant Secretary of Defense for Acquisition, memorandum for Under Secretary of Defense for Acquisition, Technology, and Logistics, May 23, 2011b.

———, *Report to Congress on the Activities of the Office of Performance Assessments and Root Cause Analyses*, Washington, D.C.: U.S. Department of Defense, 2011c, Not available to the general public.

———, "Root Cause Analysis of the Expeditionary Combat Support System Program," Washington, D.C.: Office of the Assistant Secretary of Defense for Acquisition, memorandum for Under Secretary of Defense for Acquisition, Technology, and Logistics, August 28, 2013.

———, "Root Cause Analysis of the Next Generation Operational Control System (OCX) Program," Washington, D.C.: Office of the Assistant Secretary of Defense for Acquisition, memorandum for Under Secretary of Defense for Acquisition, Technology, and Logistics, September 22, 2016.

———, "Root Cause Analysis of the Advanced Arresting Gear (AAG) Program," Washington, D.C.: Office of the Assistant Secretary of Defense for Acquisition, memorandum for Under Secretary of Defense for Acquisition, Technology, and Logistics, April 13, 2017.

———, "Root Cause Analysis of the Integrated Defensive Electronic Countermeasures Block 2/3 Subprogram," Washington, D.C.: Office of the Assistant Secretary of Defense for Acquisition, memorandum for Under Secretary of Defense for Acquisition, Technology, and Logistics, April 4, 2018.

DoD—*See* U.S. Department of Defense.

DoD CIO—*See* U.S. Department of Defense Chief Information Officer.

DoDI—*See* Department of Defense Instruction.

Drezner, Jeffrey A., Megan McKernan, Austin Lewis, Ken Munson, Devon Hill, Jaime Hastings, Geoffrey McGovern, Marek Posard, and Jerry Sollinger, *Issues with Access to Acquisition Data and Information in the Department of Defense: Identification and Characterization of Data for Acquisition Category (ACAT) II–IV, Pre-MDAPs, and Defense Business Systems*, Santa Monica, Calif.: RAND Corporation, 2019, Not available to the general public.

Eternal Sunshine of the IS Mind, "SDLC Phase 5: Maintenance," blog post, March 10, 2013. As of March 9, 2018: https://eternalsunshineoftheismind.wordpress.com/2013/03/10/sdlc-phase-5-maintenance/

Ferguson, Bob, *Leading Indicators of Program Management*, Pittsburgh, Pa.: Software Engineering Institute, Carnegie Mellon University, November 2008. As of April 22, 2019: https://pdfs.semanticscholar.org/976f/6265ea7fe39e743d893979745b4ae4617d00.pdf

Fleming, Oliver, Tim Fountaine, Nicolaus Henke, and Tamim Saleh, "Ten Red Flags Signaling Your Analytics Program Will Fail," McKinsey & Company, May 2018. As of January 10, 2018: https://www.mckinsey.com/business-functions/mckinsey-analytics/our-insights/ten-red-flags-signaling-your-analytics-program-will-fail

Freakalytics, "Tableau File Types—Purpose, Features and Limitations," webpage, August 11, 2014. As of March 13, 2019: https://www.freakalytics.com/blog/2011/08/14/tableau-file-types-purpose-features-and-limitations/

Functional Management Office, Logistics, Engineering and Force Protection, Resource Integration, "Logistics Installation Mission Support-Enterprise View (LIMS-EV)," briefing, U.S. Air Force, March 2019.

George Mason University, "Contracting with the Federal Government Programs," webpage, undated-a. As of February 13, 2018:
https://execed.gmu.edu/government-contracting/cfg-category

———, "Data Analytics Engineering, MS," webpage, undated-b. As of February 13, 2019:
https://catalog.gmu.edu/colleges-schools/engineering/data-analytics-engineering-ms/#pran

———, "Public Administration, MPA," webpage, undated-c. As of February 13, 2019:
https://catalog.gmu.edu/colleges-schools/policy-government/public-administration-mpa/#requirementstext

Georgetown University, "Master of Public Policy Curriculum," webpage, undated. As of February 13, 2019:
https://mccourt.georgetown.edu/master-in-public-policy/curriculum

George Washington University, "MS in Government Contracts Courses," webpage, undated. As of February 13, 2019:
https://business.gwu.edu/academics/programs/specialized-masters/ms-government-contracts/academic-program/courses

Georgia Institute of Technology, "Acquisition and Government Contracting," webpage, undated. As of February 13, 2019:
https://pe.gatech.edu/subjects/defense-technology/acquisition-government-contracting

GMU—*See* George Mason University.

Hardesty, Larry, "Privacy Challenges Analysis: It's Surprisingly Easy to Identify Individuals from Credit-Card Metadata," *MIT News*, January 29, 2015. As of May 4, 2019:
http://news.mit.edu/2015/identify-from-credit-card-metadata-0129

IBM Watson Studio, "Anonymizing Data for Data Protection Policies," webpage, May 1, 2019. As of May 4, 2019:
https://dataplatform.cloud.ibm.com/docs/content/wsj/governance/dmg22.html

Kay, Russell, "QuickStudy: Systems Development Life Cycle," *Computerworld*, May 14, 2002.

Kimminau, Jon A., "Five Examples of Big Data Analytics and the Future of ISR," *Joint Force Quarterly*, April 2015, pp. 30–31.

Lester, Paul B., Pedro S. Wolf, Christopher J. Nannini, Daniel C. Jensen, and Delores Johnson Davis, "Continuing the Big Data Ethics Debate: Enabling Senior Leader Decisionmaking," *Joint Force Quarterly*, April 2018, pp. 106–113.

Marine Corps Systems Command, *MARCORSYSCOM Acquisition Guidebook (MAG)*, Marine Corps Base Quantico, Va., updated February 3, 2017.

McKernan, Megan, Nancy Young Moore, Kathryn Connor, Mary E. Chenoweth, Jeffrey A. Drezner, James Dryden, Clifford A. Grammich, Judith D. Mele, Walter T. Nelson, Rebeca Orrie, Douglas Shontz, and Anita Szafran, *Issues with Access to Acquisition Data and Information in the Department of Defense: Doing Data Right in Weapon System Acquisition*, Santa Monica, Calif.: RAND Corporation, RR-1534-OSD, 2017. As of January 17, 2019: https://www.rand.org/pubs/research_reports/RR1534.html

McKernan, Megan, Jessie Riposo, Jeffrey A. Drezner, Geoffrey McGovern, Douglas Shontz, and Clifford Grammich, *Issues with Access to Acquisition Data and Information in the Department of Defense: A Closer Look at the Origins and Implementation of Controlled Unclassified Information Labels and Security Policy*, Santa Monica, Calif.: RAND Corporation, RR-1476-OSD, 2016. As of May 21, 2019: https://www.rand.org/pubs/research_reports/RR1476.html

McKernan, Megan, Jessie Riposo, Geoffrey McGovern, Douglas Shontz, and Badreddine Ahtchi, *Issues with Access to Acquisition Data and Information in the Department of Defense: Considerations for Implementing the Controlled Unclassified Information Reform Program*, Santa Monica, Calif.: RAND Corporation, RR-2221-OSD, 2018. As of May 21, 2019: https://www.rand.org/pubs/research_reports/RR2221.html

McKinsey Analytics, *Analytics Comes of Age*, New York: McKinsey & Company, January 2018. As of March 11, 2019: https://www.mckinsey.com/~/media/McKinsey/Business%20Functions/McKinsey%20Analytics/Our%20Insights/Analytics%20comes%20of%20age/Analytics-comes-of-age.ashx

MCSC—*See* Marine Corps Systems Command.

Michealson, Kirk, "Affordability Analysis: How Do We Do It?" paper presented at the Military Operations Research Society, Lockheed Martin Global Vision Center, Alexandria, Va., October 1–4, 2012. As of April 16, 2019: http://www.iceaaonline.com/ready/wp-content/uploads/2014/03/2012-MORS-Affordability-Analysis-Workshop-Report.pdf

MIT Sloan Management Review, "Lessons from Becoming a Data-Driven Organization," Massachusetts Institute of Technology, October 18, 2016. As of March 11, 2019: https://sloanreview.mit.edu/case-study/lessons-from-becoming-a-data-driven-organization/

Mosley, Mark, *DAMA-DMBOK Functional Framework*, Version 3.02, Data Management Association International, September 10, 2008. As of February 26, 2018: https://dama.org/sites/default/files/download/DAMA-DMBOK_Functional_Framework_v3_02_20080910.pdf

National Defense University, *2018–2019 Electives Program Catalog*, Washington, D.C., 2018. As of February 13, 2019: https://www.ndu.edu/Portals/59/Documents/AA_Documents/AY%2018%20-%2019%20National%20Defense%20University%20Electives%20Catalog-v3.pdf?ver=2018-08-20-113611-507

NAVAIR—*See* Naval Air Systems Command.

Naval Air Systems Command, *NAVAIR Acquisition Guide 2016/2017*, Naval Air Station Patuxent River, Md., September 22, 2015.

———, *Systems Engineering Transformation: Industry Day*, briefing, Patuxent River, Md., March 8, 2018. As of March 18, 2019: https://www.fbo.gov/utils/view?id=6e110fdff5063b075f0a99419553ef7c

Naval Postgraduate School, "NPS Data Science Certificate Program," webpage, undated-a. As of February 13, 2019: https://my.nps.edu/web/data-sciences/education-course-

———, "Operations Research (OR) Curricula," webpage, undated-b. As of February 13, 2019: https://my.nps.edu/web/or/or_curric

———, *Naval Postgraduate School Academic Catalog*, June 12, 2018. As of June 14, 2019: https://my.nps.edu/documents/104111578/106291209/Academic+Catalog+JUN18.pdf/ce7312da-97b9-490c-9b12-4529fb3dcc99

NDU—*See* National Defense University.

NPS—*See* Naval Postgraduate School.

Nott, Chris, "A Maturity Model for Big Data and Analytics," *IBM Big Data & Analytics Hub*, May 26, 2015. As of February 26, 2019: https://www.ibmbigdatahub.com/blog/maturity-model-big-data-and-analytics

Office of Management and Budget, "Federal Data Strategy, Leveraging Data as a Strategic Asset: What Are the Practices?" Office of Science and Technology Policy, Department of Commerce, Small Business Administration, webpage, undated. As of February 26, 2018: https://strategy.data.gov/practices

OPM—*See* U.S. Office of Personnel Management.

O'Rourke, Ronald, *Navy DDG-51 and DDG-1000 Destroyer Programs: Background and Issues for Congress*, Washington, D.C.: Congressional Research Service, October 22, 2018.

PARCA—*See* Director of Performance Assessments and Root Cause Analyses.

Parker, William, *Defense Acquisition University Program Managers Tool Kit*, 16th ed., Fort Belvoir, Va.: Defense Acquisition University, January 2011.

Pernin, Christopher G., Elliot Axelband, Jeffrey A. Drezner, Brian B. Dille, John Gordon IV, Bruce J. Held, K. Scott McMahon, Walter L. Perry, Christopher Rizzi, Akhil R. Shah, Peter A. Wilson, and Jerry M. Sollinger, *Lessons from the Army's Future Combat Systems Program*, Santa Monica, Calif.: RAND Corporation, RR-1206-A, 2012. As of March 12, 2019:
https://www.rand.org/pubs/monographs/MG1206.html

Public Law 111-23, Weapon Systems Acquisition Reform Act of 2009, May 22, 2009. As of June 3, 2019:
https://www.govinfo.gov/app/details/PLAW-111publ23

Public Law 114–92, National Defense Authorization Act for Fiscal Year 2016, November 25, 2015. As of May 28, 2019:
https://www.govinfo.gov/app/details/PLAW-114publ92

Public Law 114-328, National Defense Authorization Act for Fiscal Year 2017, December 23, 2016. As of May 28, 2019:
https://www.govinfo.gov/app/details/PLAW-114publ328

Public Law 115-91, National Defense Authorization Act for Fiscal Year 2018, December 12, 2017. As of May 28, 2019:
https://www.govinfo.gov/app/details/PLAW-115publ91

Riposo, Jessie, Megan McKernan, Jeffrey A. Drezner, Geoffrey McGovern, Daniel Tremblay, Jason Kumar, and Jerry Sollinger, *Issues with Access to Acquisition Data and Information in the Department of Defense: Policy and Practice*, Santa Monica, Calif.: RAND Corporation, RR-880-OSD, 2015. As of March 12, 2019:
https://www.rand.org/pubs/research_reports/RR880.html

Rollins, John, "Why We Need a Methodology for Data Science," *IBM Analytics*, August 24, 2015.

Rubinstein, Ira, "Identifiability: Policy and Practical Solutions for Anonymisation and Pseudonymisation," paper presented at the Brussels Privacy Symposium, 2016. As of May 4, 2019:
https://fpf.org/wp-content/uploads/2016/11/Rubinstein_framing-paper.pdf

Schehl, Matthew, "Naval Postgraduate School Launches Interdisciplinary Data Science and Analytics Group," Naval Postgraduate School, July 2, 2018. As of February 13, 2019: https://my.nps.edu/-/nps-launches-interdisciplinary-data-science-and-analytics-group

Smith, Gregory K., "Army AVWG Update," briefing, Washington, D.C.: U.S. Department of the Army, Assistant Secretary of the Army (Acquisition, Logistics and Technology), Acquisition Domain Functional Office, February 19, 2019, Not available to the general public.

Svensson, Jan D., "Approaching Data as an Enterprise Asset," Mastech InfoTrellis, February 26, 2013. As of May 5, 2019: http://www.infotrellis.com/approaching-data-as-an-enterprise-asset/

Taylor, Scott, "5 Master Data Best Practices," Dun & Bradstreet, June 2, 2016. As of May 5, 2016: https://www.dnb.com/perspectives/master-data/five-best-practices-master-data-management.html

Terrizzano, Ignacio, Peter Schwarz, Mary Roth, and John E. Colino, "Data Wrangling: The Challenging Journey from the Wild to the Lake," paper presented at 7th Biennial Conference on Innovative Data Systems Research, Asilomar, Calif., January 4–7, 2015. As of February 14, 2019: http://cidrdb.org/cidr2015/Papers/CIDR15_Paper2.pdf

Under Secretary of Defense (Comptroller), "DoD Budget Request," webpage, undated. As of March 29, 2019: https://comptroller.defense.gov/Budget-Materials/

———, "General Information," in *Financial Management Regulation*, Vol. 2A: *Budget Formulation and Presentation (Chapters 1–3)*, Washington, D.C.: U.S. Department of Defense, DoD 7000.14-R, October 2008. As of March 29, 2019: https://comptroller.defense.gov/Portals/45/documents/fmr/Volume_02a.pdf

———, *Financial Management Regulation*, Vol. 2B: *Budget Formulation and Presentation (Chapters 4–19)*, Washington, D.C.: U.S. Department of Defense, DoD 7000.14-R, November 2017. As of March 11, 2019: https://comptroller.defense.gov/Portals/45/documents/fmr/Volume_02b.pdf

———, *RDT&E Programs (R-1), Department of Defense Budget, Fiscal Year 2019*, Washington, D.C.: U.S. Department of Defense, revised February 13, 2018a.

———, *Department of Defense Fiscal Year (FY) 2019 Budget Estimates, Defense Advanced Research Projects Agency, Defense-Wide Justification Book*, Vol. 1 of 5: *Research, Development, Test & Evaluation, Defense-Wide*, Washington, D.C.: U.S. Department of Defense, February 2018b.

———, *Department of Defense Fiscal Year (FY) 2019 Budget Estimates, Missile Defense Agency, Defense-Wide Justification Book*, Vol. 2 of 5: *Research, Development, Test & Evaluation, Defense-Wide*, Washington, D.C.: U.S. Department of Defense, February 2018c.

———, *Department of Defense Fiscal Year (FY) 2019 Budget Estimates, Office of the Secretary of Defense, Defense-Wide Justification Book*, Vol. 3a of 5: *Research, Development, Test & Evaluation, Defense-Wide*, Washington, D.C.: U.S. Department of Defense, February 2018d.

———, *Department of Defense Fiscal Year (FY) 2019 Budget Estimates, Office of the Secretary of Defense, Defense-Wide Justification Book*, Vol. 3b of 5: *Research, Development, Test & Evaluation, Defense-Wide*, Washington, D.C.: U.S. Department of Defense, February 2018e.

———, *Department of Defense Fiscal Year (FY) 2019 Budget Estimates, Chemical and Biological Defense Program, Defense-Wide Justification Book*, Vol. 4 of 5: *Research, Development, Test & Evaluation, Defense-Wide*, Washington, D.C.: U.S. Department of Defense, February 2018f.

———, *Department of Defense Fiscal Year (FY) 2019 Budget Estimates, Defense-Wide, Defense-Wide Justification Book*, Vol. 5 of 5: *Research, Development, Test & Evaluation, Defense-Wide*, Washington, D.C.: U.S. Department of Defense, February 2018g.

———, *Fiscal Year (FY) 2019 Presidents Budget, Office of the Secretary of Defense, Defense-Wide Justification Book*, Vol. 3A of 5: *Research, Development, Test & Evaluation, Defense-Wide Budget Activities 1–3*, Washington, D.C.: U.S. Department of Defense, February 2018h.

Under Secretary of Defense for Acquisition, Technology, and Logistics, *Performance of the Defense Acquisition System: 2013 Annual Report*, Washington, D.C.: U.S. Department of Defense, 2013. As of March 9, 2018: http://www.dtic.mil/docs/citations/ADA587235

———, *Performance of the Defense Acquisition System: 2014 Annual Report*, Washington, D.C.: U.S. Department of Defense, 2014. As of March 9, 2018: http://www.dtic.mil/docs/citations/ADA603782

———, *Performance of the Defense Acquisition System: 2015 Annual Report*, Washington, D.C.: U.S. Department of Defense, 2015. As of March 9, 2018: http://www.dtic.mil/docs/citations/ADA621941

———, *Performance of the Defense Acquisition System: 2016 Annual Report*, Washington, D.C.: U.S. Department of Defense, October 2016a. As of March 9, 2018: http://www.dtic.mil/docs/citations/AD1019605

———, *Department of Defense Acquisition Workforce Strategic Plan: FY 2016–FY 2021*, Washington, D.C.: U.S. Department of Defense, December 2016b. As of March 18, 2018: http://www.hci.mil/docs/Policy/Legal%20Authorities/DoD_Acq_Workforce_Strat_Plan_FY16_FY21.pdf

United States Code, Title 10, Armed Forces. As of May 28, 2019: https://www.govinfo.gov/app/details/USCODE-2011-title10/

United States Code, Title 10, Section 2222, Defense Business Systems: Architecture, Accountability, and Modernization. As of May 28, 2019: https://www.govinfo.gov/app/details/USCODE-2011-title10/USCODE-2011-title10-subtitleA-partIV-chap131-sec2222

United States Code, Title 10, Section 2302, Definitions. As of May 28, 2019: https://www.govinfo.gov/app/details/USCODE-2017-title10/USCODE-2017-title10-subtitleA-partIV-chap137-sec2302

United States Code, Title 10, Section 2440, Technology and Industrial Base Plans. As of May 28, 2019: https://www.govinfo.gov/app/details/USCODE-2017-title10/USCODE-2017-title10-subtitleA-partIV-chap144-sec2440

U.S. Bureau of Labor Statistics, Division of Occupational Employment Statistics, "May 2016 National Occupational Employment and Wage Estimates—United States," webpage, updated May 2017. As of February 28, 2018: https://www.bls.gov/oes/current/oes_nat.htm

U.S. Department of Defense, Chief Information Officer, Resources and Analysis, *Department of Defense Information Technology Budget Exhibit: Fiscal Year 2013 President's Budget Request, IT-1 Report*, Washington, D.C.: U.S. Department of Defense, March 2013. As of March 29, 2019: https://www.cape.osd.mil/content/SNAPIT/BudgetDocs2013.html

———, *Department of Defense Information Technology Budget Exhibit: Fiscal Year 2015 President's Budget Request, IT-1 Report*, Washington, D.C.: U.S. Department of Defense, April 2014. As of March 29, 2019: https://www.cape.osd.mil/content/SNAPIT/BudgetDocs2015.html

———, *Department of Defense Information Technology Budget Exhibit: Fiscal Year 2018 President's Budget Request, IT-1 Report*, Washington, D.C.: U.S. Department of Defense, March 2017.

———, *Department of Defense Information Technology Budget Exhibit: Fiscal Year 2019 President's Budget Request, IT-1 Report*, Washington, D.C.: U.S. Department of Defense, March 2018. As of March 29, 2019: https://www.cape.osd.mil/content/SNAPIT/BudgetDocs2019.html

U.S. House of Representatives, *National Defense Authorization Act for Fiscal Year 2017: Conference Report to Accompany S. 2943*, Washington, D.C., Report 114-840, November 30, 2016. As of May 23, 2019: https://www.congress.gov/congressional-report/114th-congress/house-report/840

U.S. Office of Management and Budget, "Federal Data Strategy, Leveraging Data as a Strategic Asset: What Are the Practices?" Office of Science and Technology Policy, Department of Commerce, Small Business Administration, webpage, undated. As of February 26, 2018: https://strategy.data.gov/practices

U.S. Office of Personnel Management, "Classification and Qualifications: Classifying General Schedule Positions," webpage, undated-a. As of February 22, 2018: https://www.opm.gov/policy-data-oversight/classification-qualifications/classifying-general-schedule-positions/

———, "FedScope," webpage, undated-b. As of December 10, 2018: https://www.fedscope.opm.gov/

———, *Handbook of Occupational Groups and Families*, Washington D.C., May 2009. As of February 22, 2018: https://www.opm.gov/policy-data-oversight/classification-qualifications/classifying-general-schedule-positions/occupationalhandbook.pdf

U.S. Small Business Administration, "Frequently Asked Questions: General Questions," webpage, undated. As of March 11, 2019: https://www.sbir.gov/faqs/general-questions

vander Meulen, Rob, and Thomas McCall, "Gartner Survey Shows Organizations Are Slow to Advance in Data and Analytics," press release, Gartner, Inc., February 5, 2018. As of February 26, 2018: https://www.gartner.com/en/newsroom/press-releases/2018-02-05-gartner-survey-shows-organizations-are-slow-to-advance-in-data-and-analytics

Work, Robert O., "The Littoral Combat Ship: How We Got Here, and Why," white paper, U.S. Navy, 2013. As of March 12, 2019: https://awin.aviationweek.com/Portals/AWeek/Ares/work%20white%20paper.PDF